Photonic Imaging for Biology

SCIENCES

Image, Field Director – Laure Blanc-Féraud

Imagery in Life Sciences, Subject Head – Françoise Peyrin

Photonic Imaging for Biology

From Conventional Microscopy to Super-Resolution

Coordinated by

Jean-Baptiste Sibarita

iSTE

WILEY

First published 2025 in Great Britain and the United States by ISTE Ltd and John Wiley & Sons, Inc.

ISTE Ltd
27-37 St George's Road
London SW19 4EU
UK

www.iste.co.uk

John Wiley & Sons, Inc.
111 River Street
Hoboken, NJ 07030
USA

www.wiley.com

Library of Congress Control Number: 2025936520

British Library Cataloguing-in-Publication Data
A CIP record for this book is available from the British Library
ISBN 978-1-78945-222-8

ERC code:
LS3 Cell Biology, Development, Stem Cells and Regeneration
LS3_16 Functional imaging of cells and tissues

The manufacturer's authorized representative according to the EU General Product Safety Regulation is Wiley-VCH GmbH, Boschstr. 12, 69469 Weinheim, Germany, e-mail: Product_Safety@wiley.com.

Contents

Preface

Photonic Microscopy for Biology: From Microscopy to Super-Resolution

Jean-Baptiste SIBARITA
Interdisciplinary Institute for Neuroscience, UMR 5297, CNRS, University of Bordeaux, France

Light microscopy is one of the most fundamental and widely used tools in the biological sciences, enabling the monitoring of living cells and organisms in their native environment, with details invisible to the naked eye. It employs visible light and a series of optical components to magnify and resolve fine biological structures and their dynamics, revealing the innermost hidden secrets of life.

Light microscopy has undergone several key developments since its inception thanks to scientists' ingenuity and creativity. From the first scientific microscopes in the 17th century, advances in optics (Carl Zeiss and Ernst Abbe) and staining (Gram) in the late 19th century, microscopy has witnessed transformative changes in the 20th century with Fritz Zernike inventing phase contrast microscopy, Albert Coons developing immunofluorescence, Marvin Minsky introducing the confocal microscope and George Nomarski pioneering differential interference contrast. The discovery and application of green fluorescent protein (GFP) by Osamu Shimomura, Martin Chalfie and Roger Tsien, who were awarded the Nobel Prize in Chemistry in 2008, has emerged as one of the most important tools in modern cell and molecular biology. Recent breakthroughs in the 21st century include super-resolution microscopy (Eric Betzig, Stefan W. Hell and William E. Moerner, who were awarded the Nobel Prize in Chemistry in 2014) and artificial intelligence

(AI), which now plays a crucial role in image analysis, enhancing the interpretation of complex datasets.

Modern instruments incorporate advanced optics and electronics, digital imaging and computerized methods. Fluorescence enables the visualization of biomolecules with high specificity and contrast, and super-resolution offers near-molecular resolution.

Today, different types of microscopy cater to various scientific needs, enabling breakthroughs in understanding life at multiple scales. This book aims to present these various microscopy techniques, their principles, advantages and limits. Inevitably, it is such a vast and prolific field that it is impossible to cover all of them.

This book emphasizes approaches that have contributed the most to discoveries in biology, explained by experts in the field. It starts with fundamental principles of optical microscopy, introducing image formation, resolution and optical sectioning (Chapter 1), as well as brightfield microscopy techniques such as phase contrast and differential interference contrast (DIC) (Chapter 2). The book then continues with a chapter on fluorescence, describing fluorophore photophysics, fluorescent proteins and organic fluorophores (Chapter 3), followed by a repertoire of fluorescence-based techniques, covering fluorescence recovery after photobleaching and fluorescence correlation spectroscopy (FRAP and FCS) (Chapter 4), single particle tracking (SPT) (Chapter 5) and in-depth fluorescence microscopy, including scanning and spinning-disk confocal, multi-photons and light-sheet microscopy (Chapter 6). It follows with the more recent super-resolution techniques: structured illumination microscopy (SIM) (Chapter 7), stimulated emission depletion (STED) (Chapter 8) and single molecule localization microscopy (SMLM) (Chapter 9). The book concludes with a chapter on image processing and analysis, including recent advancements in deep learning approaches (Chapter 10).

All of these techniques, at the interface of various disciplines, have contributed to revolutionize cell biology, neuroscience and developmental biology, always pushing the boundaries of what can be visualized and quantified in living systems further.

I would like to thank all of the authors for their work, their time and their patience, and for agreeing to share their expertise in this book.

August 2025

1

Principles of Light Microscopy

Guillaume DUPUIS

Institut des Sciences Moléculaires d'Orsay, Université Paris-Saclay, France

1.1. Introduction

Since its invention in the 17th century, the optical microscope has become an essential instrument widely used in many fields of application, from metallurgy to biology and medicine. More specifically, in the field of life sciences, despite the considerable progress achieved by electron microscopes and local probe microscopes, the broadly preferred imaging techniques are still, at the current time, optical techniques that make use of combinations of conventional lenses and mirrors, as well as visible light sources. This is because cells, as the preferred objects of study in the life sciences, are indeed optically transparent: only light microscopy provides a three-dimensional noninvasive imaging of the inside of cells, under conditions compatible with living organisms.

In addition, mainly through the use of various fluorescent labeling techniques, light microscopy allows for the detection of specific cell constituents, such as proteins or lipids. Fluorescence light microscopy would almost be the ideal imaging technique for studying biological objects if it were able to distinguish between details smaller than about 200 nm. Unfortunately, it has been established since the studies of Ernst Abbe (1840–1905), a German optician, that the spatial resolution of any optical instrument is intrinsically limited by the diffraction of light.

In fact, due to the diffraction of light by the limited aperture of a lens, a perfect source point cannot yield an equally point image, but merely a diffraction pattern that is all the larger when the aperture is small and the desired magnification is high. Conversely, a large aperture will result in a shallow depth of field: this quickly becomes a problem for thick samples, especially in fluorescence microscopy, when the fluorescence from out-of-focus planes degrades the contrast of the plane of focus.

The main developments therefore consist of the following:

– improving the contrast of specific objects, especially thick samples, by implementing optical sectioning techniques such as, in particular, confocal microscopy, structured illumination techniques, light sheet illumination and total internal reflection microscopy;

– implementing strategies for overcoming the diffraction limit using super-resolution techniques.

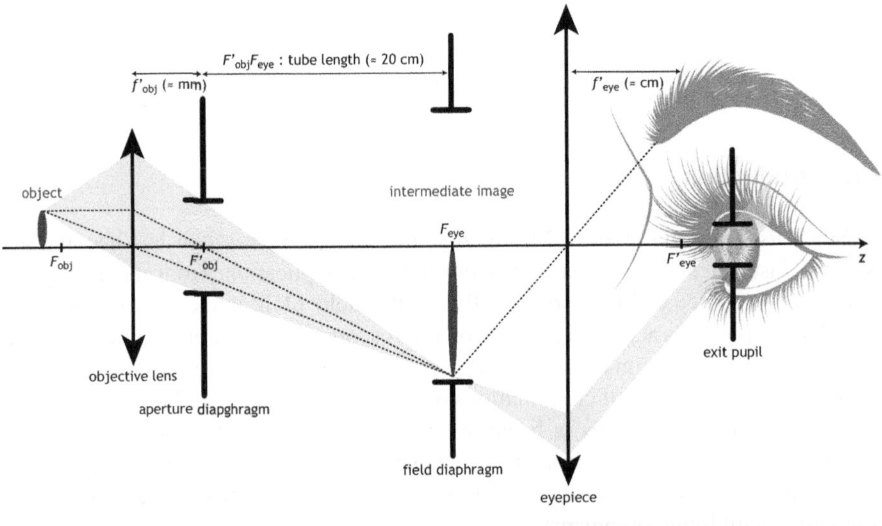

Figure 1.1. *Schematic diagram of a compound microscope. The magnified image of the object is formed by the objective lens whose focal length f'_{obj} is small (about a few millimeters). The object is placed in front of and at a short distance from the object focal plane of the objective F_{obj} so that the absolute value of the magnification is large. The aperture diaphragm is placed in the back focal plane F'_{obj} of the lens to provide telecentricity in object space. The intermediate image is formed in the object focal plane F_{eye} of the eyepiece and the final image is projected to infinity. For a color version of this figure, see www.iste.co.uk/sibarita/photonic.zip*

1.2. Principle of image formation

1.2.1. *Geometric approach*

In a traditional compound microscope such as the one shown in Figure 1.1, the magnified image of the object is formed by a first optical group called the *objective lens*. The objective lens is the crucial element of the microscope that usually determines the overall performance of the instrument. The object to be observed is placed in the vicinity of the object focal plane F_{obj} of the lens. The magnified image is formed in an image plane located well beyond the image focus plane F'_{obj} of the lens. Obviously, the magnification will be greater the shorter the object focal length of the lens and the shorter the object plane. The magnified image formed by the objective lens is real and inverted and it traverses a second optical group called the *eyepiece*. The eyepiece is positioned so that the intermediate image falls within its focal object plane F_{eye}. Thereby, the eye observes an image projected to infinity (for a standard observer), thus relaxing the muscles responsible for accommodation, resulting in better visual comfort. Eyepieces are generally characterized by their commercial magnification, i.e. the ratio between the angle from which the final image is viewed and the angle from which the object would be seen if it were placed at the *least distance of distinct vision* of the eye (at 250 mm). The commercial magnification of the global microscope is therefore the product between the magnification of the objective and the commercial magnification of the eyepiece.

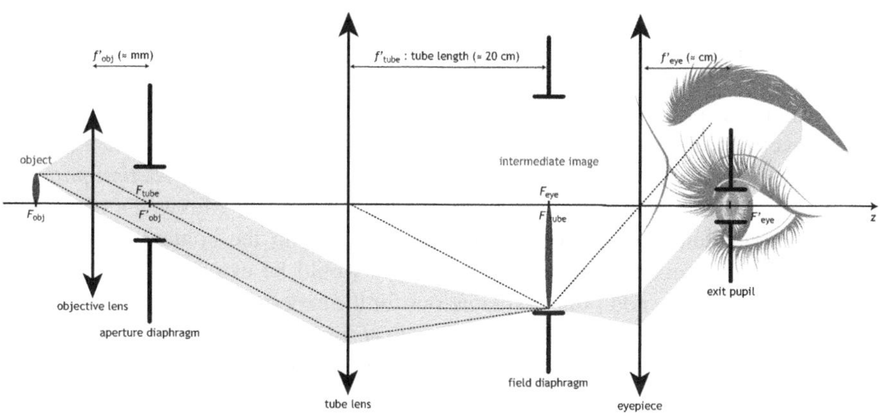

Figure 1.2. *Schematic diagram of a microscope using an infinity-corrected objective lens. The object is placed in the object focal plane F_{obj} of the lens, which is associated with a tube lens with a focal length of about 20 cm. For a color version of this figure, see www.iste.co.uk/sibarita/photonic.zip*

In microscopes that use "*infinity-corrected*" optics (in the sense of aberration correction) such as the one represented schematically in Figure 1.2, the object is in the object focal plane of the objective lens, which consequently provides an image projected to infinity. An additional lens called *a tube lens* reforms the intermediate image in the object focal plane of the eyepiece. The object focus plane of the tube lens is coincident with the back focal plane of the objective lens so that the whole forms an afocal conjugation whose magnification is that indicated on the objective lens. In this type of configuration, the eyepiece circle, i.e. the exit pupil of the microscope, is located in the image focal plane of the eyepiece: the size of the image is then independent of the instrument setting. If the sample is observed with a matrix detector, the matrix detector is placed in the image focal plane of the tube lens or in a conjugate plane therewith.

The aperture diaphragm of the microscope is placed in the back focal plane of the lens so that the system is telecentric in object space. The main rays (namely the oblique rays that travel through the center of the aperture diaphragm) are then parallel to the optical axis in the object space, thus producing an orthographic view of the sample: the magnification is independent of the position of the sample in the field.

1.2.2. *Wave approach*

Geometric optics do not take into account the wave character of light. Nonetheless, diffraction phenomena are central to the process of image formation in a microscope, particularly in terms of the spatial resolution of the images.

Diffraction is a general characteristic of wave phenomena and occurs when a portion of the wavefront meets an obstacle: from the moment a wave is altered, whether in phase or amplitude, due to an obstacle, it is diffracted. The different parts of the wavefront that propagate beyond the obstacle recombine and interfere with one another, resulting in an illumination distribution that is called a *diffraction pattern*. Diffraction can be studied within the very general context of Maxwell's equations for the propagation of electromagnetic waves or even in the context of quantum electrodynamics. Nevertheless, these approaches would be extremely cumbersome and would provide few additional explanations for the experimental observations. The classical scalar wave theory used in the *Huygens–Fresnel principle* provides a simple and efficient formalism that proves to be more than sufficient.

1.2.2.1. *The Huygens–Fresnel principle*

The theory of diffraction, in its elementary form, is based on a principle proposed by Christiaan Huygens (1629–1695) in his *Treatise on Light*. According to this principle, each point reached by a wave behaves as a secondary source of spherical wavelets, so that a moment later, the resulting wavefront is the envelope of those wavelets. Another crucial point of his proposal is that this propagation occurs without any change in wavelength or wave velocity, provided that there is no change in medium. Augustin Jean Fresnel (1788–1827), in the 1800s, modified this Huygens principle to successfully introduce the concept of interference between secondary wavelets. The combined contribution of these two physicists is called the *Huygens–Fresnel principle*.

About 50 years later, Gustav Robert Kirchhoff (1824–1887) showed that the Huygens–Fresnel principle was a direct consequence of the scalar differential equation of wave propagation. One hundred years later, it was Arnold Sommerfeld (1868–1951) who reproved the whole Huygens–Fresnel principle in the more rigorous context of vector electromagnetism, i.e. of the solution of Maxwell's vector equations. In summary, the Huygens–Fresnel principle is an approximation of the rigorous solution to the diffraction problem given by solving Maxwell's equations. It is valid in the context of the *scalar* approximation, i.e. in the context of the *paraxial* approximation. The rays are therefore not very far away and slightly sloped with respect to the optical axis (typically, less than $\sim 18°$), and the distance between the object and the diffraction pattern is large in comparison to both the size of the object and the size of the diffraction pattern.

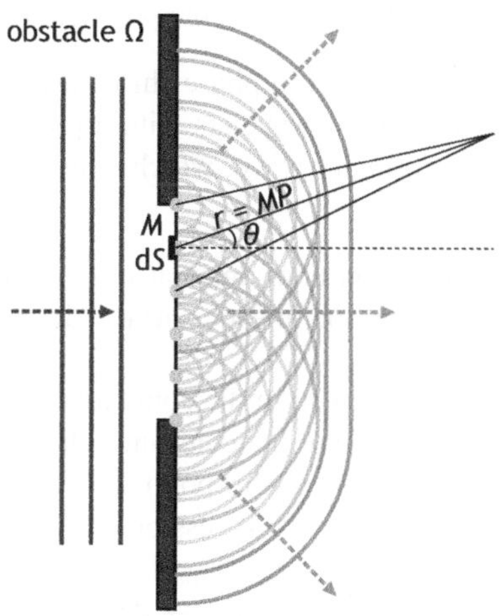

Figure 1.3. *Diffraction of a plane wave on an obstacle according to the Huygens–Fresnel principle. For a color version of this figure, see www.iste.co.uk/sibarita/photonic.zip*

NOTATIONS.– The obstacle is denoted as Ω, M is a point of the diffracting aperture, $\mathrm{d}S$ is a surface element of the diffracting aperture centered at M, P is the point where the amplitude of the diffracted wave is to be expressed, r is the distance MP in the context of the scalar approximation, and θ is the angle of obliquity of MP with respect to the normal at the diffracting aperture.

FORMULATION.– The notations are shown in Figures 1.3 and 1.4. Let a monochromatic wave be incident on an aperture Ω. Each unobstructed point of the wavefront, at a given instant, can be considered as a secondary source of spherical wavelets of the same wavelength as the incident wave (Huygens). For any point located beyond the aperture, the scalar optical amplitude is the superposition in amplitude and phase (Fresnel) of all of the spherical wavelets: the wavelets interfere with each other to yield the globally diffracted wave. The literal mathematical translation of the Huygens–Fresnel principle allows the amplitude of the wave at the point P to be expressed as:

$$\mathcal{A}(P) = \iint_\Omega \mathcal{A}(M)\, K(\theta)\, \frac{\mathrm{e}^{\,jkr}}{r}\, \mathrm{d}S \quad \text{with} \quad K(\theta) = -\frac{j}{\lambda}\frac{(1+\cos\theta)}{2} \qquad [1.1]$$

where $k = 2\pi/\lambda$ is the wave vector and $K(\theta)$ is called the *obliquity factor*. Without going into too much detail about the calculation, we try to explain the origin of the different terms of this expression:

– The angular dependence leads to canceling out $K(\theta)$ for backward propagation ($\theta = \pi$), which follows the natural direction of the propagation of light.

– The term $-j = \mathrm{e}^{\,-j\pi/2}$ is, in fact, a phase shift[1]: everything happens as if the secondary sources were emitting their wavelets with a phase shift of a quarter of a wavelength with respect to the primary wave, which is in agreement with the theory of electromagnetic dipole radiation.

– The term $1/\lambda$ is an integration constant that ensures the homogeneity of the expression and weights the contribution of each surface element $\mathrm{d}S$.

FRESNEL APPROXIMATION.– Moreover, in the context of scalar approximation, it is possible to consider the angle relative to the constant obliquity factor around a value θ_0 only as a function of the position of the point P. We can therefore factor the obliquity coefficient within the integral of expression [1.1] and write:

$$\mathcal{A}(P) = -\frac{j}{\lambda}\frac{(1+\cos\theta_0)}{2} \iint_\Omega \mathcal{A}(M)\, \frac{\mathrm{e}^{\,jkr}}{r}\, \mathrm{d}S \qquad \text{(scalar approximation)} \qquad [1.2]$$

Finally, in paraxial optics, the angle of view θ_0 can even be considered small enough for its cosine to be equal to 1. The relative error remains less than 5% as long as $\theta_0 < 18°$. Expression [1.2] becomes:

$$\mathcal{A}(P) = -\frac{j}{\lambda} \iint_\Omega \mathcal{A}(M)\, \frac{\mathrm{e}^{\,jkr}}{r}\, \mathrm{d}S \qquad \text{(paraxial approximation)} \qquad [1.3]$$

1. As any phase shift, the sign is conventional: if we take as a sign convention for the propagation of the wave a term in $\mathrm{e}^{\,-jkr}$, the obliquity factor appears in $+j$ and not in $-j$.

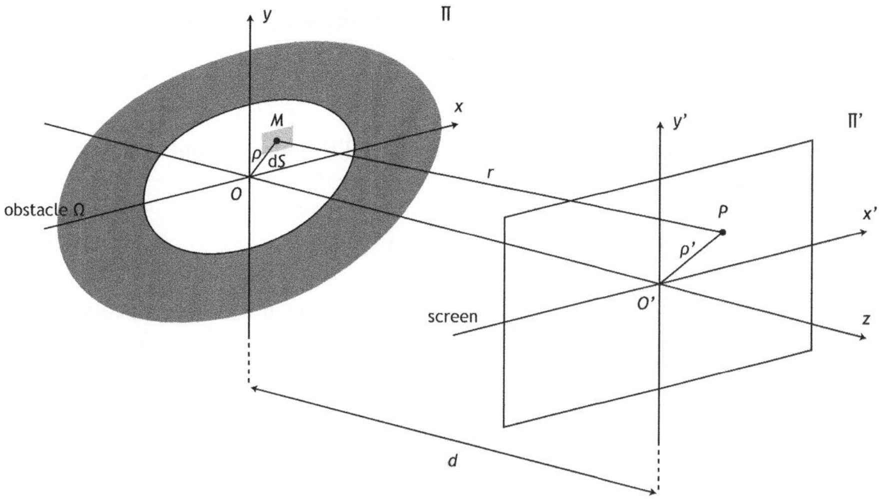

Figure 1.4. *Huygens–Fresnel principle in paraxial optics*

NOTATIONS.– The obstacle is denoted as Ω, M is a point of the diffracting aperture located in the plane Π, $\mathrm{d}S$ is a surface element of the diffracting aperture centered on M, P is a point of the plane Π' where we seek to express the amplitude of the diffracted wave, d is the distance between the planes Π and Π', assumed to be parallel and orthogonal to the optical axis (Oz), r is the distance MP in the framework of the scalar approximation, and ρ and ρ' are the radial coordinates of M and P in their respective transverse planes.

Since the aperture angle is small, the distance r varies little regardless of the exact position of M in the diffracting structure Ω. The quantity $1/r$ can therefore be factored from the integral and considered as a first approximation equal to the distance d separating the planes Π and Π'. Bearing in mind the notations used, the integral takes the following form:

$$\mathcal{A}(P) = -\frac{j}{\lambda d} \iint_{\Omega} \mathcal{A}(M)\, \mathrm{e}^{\,jkr}\, \mathrm{d}S$$

Similarly, since the phase of the exponential is proportional to r, the latter can also be replaced by a limited expansion, but not at the first order since the phase is a rapidly varying quantity. Fresnel's approximation consists of keeping only the second-order term. It follows that:

$$r = \sqrt{d^2 + (\vec{\rho} - \vec{\rho}\,')^2} \approx d \left(1 + \frac{1}{2}\frac{(\vec{\rho} - \vec{\rho}\,')^2}{d^2}\right)$$

The phase $\phi = kr$ can therefore be replaced by the simplified expression:

$$\phi = kr \approx kd\left(1 + \frac{1}{2}\frac{(\vec{\rho} - \vec{\rho}')^2}{d^2}\right) = kd + 2\pi\,\frac{(\vec{\rho} - \vec{\rho}')^2}{2\lambda d}$$

Under these conditions, the Fresnel–Kirchhoff diffraction formula can be written in the following manner, known as the *Fresnel integral*:

$$\mathcal{A}(P) = -\frac{j}{\lambda d}\,\mathrm{e}^{\,jkd}\iint_{\Omega}\mathcal{A}(M)\,\mathrm{e}^{\,2j\pi\,\frac{(\vec{\rho}-\vec{\rho}')^2}{2\lambda d}}\,\mathrm{d}S \qquad [1.4]$$

We next expand the phase of the complex exponential that appears in the expression of the Fresnel integral [1.4] with $(\vec{\rho} - \vec{\rho}')^2 = \rho'^2 + \rho^2 - 2\,\vec{\rho}\cdot\vec{\rho}'$:

$$\mathcal{A}(P) = -\frac{j}{\lambda d}\,\mathrm{e}^{\,jkd}\,\mathrm{e}^{\,j\pi\,\frac{\rho'^2}{\lambda d}}\iint_{\Omega}\left(\mathcal{A}(M)\,\mathrm{e}^{\,j\pi\,\frac{\rho^2}{\lambda d}}\right)\,\mathrm{e}^{\,-2j\pi\,\frac{\vec{\rho}\cdot\vec{\rho}'}{\lambda d}}\,\mathrm{d}S$$

however $\vec{\rho}\cdot\vec{\rho}' = xx' + yy'$ thereby $\mathrm{e}^{\,-2j\pi\,\frac{\vec{\rho}\cdot\vec{\rho}'}{\lambda d}} = \mathrm{e}^{\,-2j\pi\,\frac{xx'+yy'}{\lambda d}}$

let us set $u'\frac{x'}{\lambda d}$ and $v' = \frac{y'}{\lambda d}$ (homogeneous to lengths^{-1})

$$\mathcal{A}(P) = -\frac{j}{\lambda d}\,\mathrm{e}^{\,jkd}\,\mathrm{e}^{\,j\pi\,\frac{\rho'^2}{\lambda d}}\iint_{\Omega}\left(\mathcal{A}(M)\,\mathrm{e}^{\,j\pi\,\frac{\rho^2}{\lambda d}}\right)\,\mathrm{e}^{\,-2j\pi\,(u'x+v'y)}\,\mathrm{d}S$$

$$\mathcal{A}(P) = -\frac{j}{\lambda d}\,\mathrm{e}^{\,jkd}\,\mathrm{e}^{\,j\pi\,\frac{\rho'^2}{\lambda d}}\,\mathcal{FT}\left[\mathcal{A}(M)\,\mathrm{e}^{\,j\pi\,\frac{\rho^2}{\lambda d}}\right]\left(u' = \frac{x'}{\lambda d},\, v' = \frac{y'}{\lambda d}\right) \qquad [1.5]$$

The calculation of the complex amplitude in the plane Π' is therefore equivalent to performing a Fourier transform.

1.2.2.2. *Diffraction within a centered optical system*

The Huygens–Fresnel principle as described above does not mention any optics but merely the propagation in free space of a monochromatic plane wave through a limited aperture. We now look into what happens in a centered optical system such as a microscope, used under paraxial conditions. We consider a point source S, placed in the object plane on the axis and study the image of this source in the paraxial image plane (see Figure 1.5A).

Given that stigmatic conditions are guaranteed by the paraxial approximation, the wave that converges at point S' as an image of S by the optical system is a spherical wave centered on O' and resting on the edges of the exit pupil of the optical system. Let $M(x, y)$ be a point on the plane containing the exit pupil and $P(x', y')$ a point on the paraxial plane. The complex amplitude at point P is given by the Fresnel integral ($\rho = \sqrt{x^2 + y^2}$ and $\rho' = \sqrt{x'^2 + y'^2}$):

$$\mathcal{A}(P) = -\frac{j}{\lambda d}\,\mathrm{e}^{\,jkd}\,\mathrm{e}^{\,j\pi\,\frac{\rho'^2}{\lambda d}}\,\mathcal{FT}\left[\mathcal{A}(M)\,\mathrm{e}^{\,j\pi\,\frac{\rho^2}{\lambda d}}\right]\left(u' = \frac{x'}{\lambda d},\, v' = \frac{y'}{\lambda d}\right) \qquad [1.5]$$

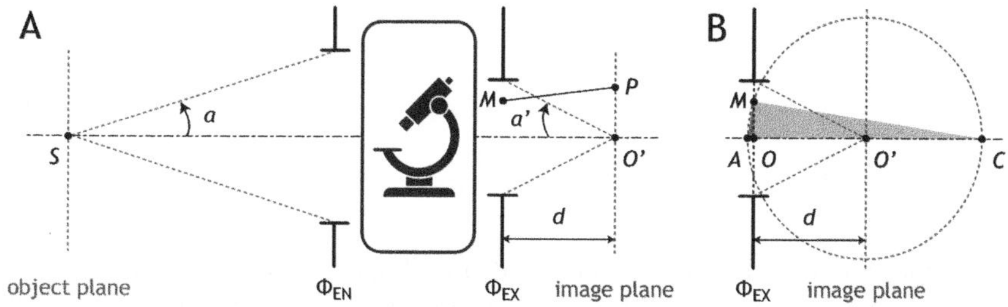

Figure 1.5. *Diffraction of the image of a point by a centered optical system, under the conditions of the paraxial approximation. A. Φ_{EN} and Φ_{EX}, respectively, represent the entrance and exit pupils of the optical system. B. Spherical wave resting on the edges of the exit pupil: illustration of the radial phase shift that exists in the plane of the exit pupil. For a color version of this figure, see www.iste.co.uk/sibarita/photonic.zip*

Should the wave be flat in the plane of the exit pupil, then $\mathcal{A}(M)$ would be constant and equal to $\mathcal{A}_0$. Yet the wave is spherical: the points M on the periphery of the pupil are found ahead of the points M located near the optical axis (see Figure 1.5B). Assuming that the point O (center of the exit pupil) is the origin of the phases, the term $\mathcal{A}(M)$ is written as:

$$\mathcal{A}(M) = \mathcal{A}_0 \, \mathrm{e}^{-j\varphi(M)}$$

with $\varphi(M) = \varphi(M) - \varphi(O) = \varphi(A) - \varphi(O) = \dfrac{2\pi}{\lambda} OA$

As can be seen in Figure 1.5B, the geometry of the problem reveals three right-angled triangles: (A, O, M) and (M, O, C) rectangles at O, and (A, M, C) rectangle at M. Based on the Pythagorean theorem, it can be written that:

$$AM^2 = OA^2 + OM^2 \qquad MC^2 = OM^2 + OC^2$$

$$AC^2 = AM^2 + MC^2 = (OA + OC)^2$$

$$AM^2 + MC^2 = OA^2 + 2\,OA \times OC + OC^2 = OA^2 + 2OM^2 + OC^2$$

$$OA = \frac{OM^2}{OC} \approx \frac{OM^2}{2d} = \frac{\rho^2}{2d}$$

Taking into account the finite aperture of the exit pupil, represented by the pupil function $\Omega(M)$, the amplitude in the plane of the exit pupil is therefore written as:

$$\mathcal{A}(M) = \mathcal{A}_0 \, \Omega(M) \, \mathrm{e}^{-j\pi \frac{\rho^2}{\lambda d}} \qquad [1.6]$$

In the image plane, according to expression [1.5] of the Fresnel integral, the complex amplitude is written as:

$$\mathcal{A}(P) = -\frac{j}{\lambda d} \, \mathrm{e}^{jkd} \, \mathrm{e}^{j\pi \frac{\rho'^2}{\lambda d}} \, \mathcal{FT}\left[\mathcal{A}(M) \, \mathrm{e}^{j\pi \frac{\rho^2}{\lambda d}}\right] \left(u' = \frac{x'}{\lambda d}, \, v' = \frac{y'}{\lambda d}\right) \qquad [1.5]$$

however

$$\mathcal{A}(M) = \mathcal{A}_0\,\Omega(M)\,\mathrm{e}^{-j\pi\frac{\rho^2}{\lambda d}} \quad [1.6]$$

$$\Rightarrow \quad \mathcal{A}(P) = -\frac{j}{\lambda d}\,\mathrm{e}^{jkd}\,\mathrm{e}^{j\pi\frac{\rho'^2}{\lambda d}}\,\mathcal{FT}\left[\mathcal{A}_0\,\Omega(M)\,\mathrm{e}^{-j\pi\frac{\rho^2}{\lambda d}}\,\mathrm{e}^{j\pi\frac{\rho^2}{\lambda d}}\right]$$

For the record, in the argument of the Fourier transform, the term in positive exponential originates from the divergent aspect of the wavelets of the Huygens–Fresnel principle, while the term in negative exponential originates from the convergence of the spherical wave from the exit pupil to the image plane. The two terms cancel out one another perfectly.

$$\Rightarrow \quad \mathcal{A}(P) = -\mathcal{A}_0\,\frac{j}{\lambda d}\,\mathrm{e}^{jkd}\,\mathrm{e}^{j\pi\frac{\rho'^2}{\lambda d}}\,\mathcal{FT}\left[\Omega(M)\right]\left(u' = \frac{x'}{\lambda d},\, v' = \frac{y'}{\lambda d}\right) \quad [1.7]$$

Irradiance is therefore written as:

$$\Rightarrow \quad I(P) = \mathcal{A}(P)\,\mathcal{A}^*(P) = \frac{|\mathcal{A}_0|^2}{(\lambda d)^2}\left|\mathcal{FT}\left[\Omega(M)\right]\left(u' = \frac{x'}{\lambda d},\, v' = \frac{y'}{\lambda d}\right)\right|^2 \quad [1.8]$$

It should be observed that for a point source on the optical axis in the object plane, the diffraction pattern in the conjugate image plane is given by the Fourier transform of the pupil function, taken at spatial frequencies $u' = x'/\lambda d$ and $v' = y'/\lambda d$. The light amplitude distribution in the image plane for a point source on the axis in the object plane given by expression [1.7] is called *percussion response in coherent lighting*[2]. If a matrix detector is placed in the image plane, it records the corresponding irradiance, i.e. the square modulus of the percussion response. Once normalized, this irradiance distribution function in the image plane for a point source is called *Point Spread Function* (PSF).

1.2.2.3. *Circular pupil and Airy disk*

The circular pupil is the most common case in microscopy. The diffraction pattern resulting from light traveling through a circular hole is called the *Airy disk*. The name of this pattern was coined by George Biddell Airy (1801–1892), an English scientist who discovered and described the phenomenon in 1835 in *On the Diffraction of an Object-glass with Circular Aperture*.

2. The designation "*percussion response*" is used to reflect the fact that the optical system response affects a two-dimensional space in a noncausal way. When a system depends on only one variable (time, for example), the designation "*impulse response*" is used because of the one-dimensional and causal nature of this response.

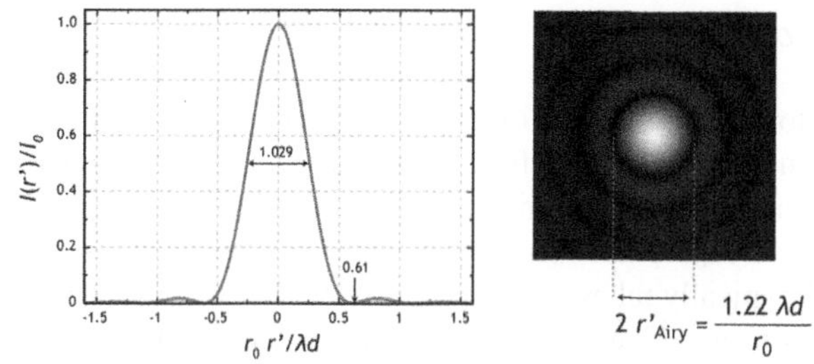

Figure 1.6. *Diffraction pattern associated with a circular pupil: Airy disk. For a color version of this figure, see www.iste.co.uk/sibarita/photonic.zip*

Without delving too far into the mathematical demonstration, the transmission factor of a pupil with radius r_0 is written as follows:

$$\Omega(r) = \text{disc}\left(\frac{r}{r_0}\right) \quad \text{with} \quad r = \sqrt{x^2 + y^2} \qquad [1.9]$$

According to expression [1.7], the complex light amplitude $\mathcal{A}(r')$ in the image plane is proportional to the Fourier transform of the diffracting aperture, taken at the spatial frequency $w' = r'/\lambda d = \sqrt{x'^2 + y'^2}/\lambda d$:

$$\mathcal{A}(r') = -\mathcal{A}_0 \frac{j}{\lambda d} \mathrm{e}^{jkd} \mathrm{e}^{j\pi \frac{\rho'^2}{\lambda d}} \mathcal{FT}\left[\Omega(r)\right]\left(w' = \frac{r'}{\lambda d}\right)$$

$$\mathcal{A}(r') = -\mathcal{A}_0 \frac{j}{\lambda d} \mathrm{e}^{jkd} \mathrm{e}^{j\pi \frac{\rho'^2}{\lambda d}} \pi {r_0}^2 \frac{2\, J_1(2\pi r_0\, r'/\lambda d)}{2\pi r_0\, r'/\lambda d} \qquad [1.10]$$

The corresponding irradiance is thus:

$$I(r') = \mathcal{A}(r')\mathcal{A}^*(r') = \frac{1}{(\lambda d)^2} \mathcal{A}_0^2 \left(\pi {r_0}^2\right)^2 \left|\frac{2\, J_1(2\pi r_0\, r'/\lambda d)}{2\pi r_0\, r'/\lambda d}\right|^2 \qquad [1.11]$$

The function J_1 is the first-order Bessel function of the first kind. In expressions [1.10] and [1.11], the term $2\, J_1(2\pi r_0\, w')/(2\pi r_0\, w')$ must be considered as the analogue of a cardinal sine for a two-dimensional system with symmetry of revolution, with the difference being that the function J_1 exhibits oscillations that have neither the period nor the amplitude of a sinusoid.

Similarly to the pupil that generates it, the diffraction pattern exhibits a symmetry of revolution and takes the form of a bright spot surrounded by halos of concentric circles of lower luminosity. The first zero of the function J_1 is located at an argument equal to $3.83 \approx 1.22\pi$ (instead of π for the cardinal sine function). The radius of the Airy disk is:

$$r'_{\text{Airy}} = 0.61 \frac{\lambda d}{r_0} = 1.22 \frac{\lambda}{2\,\text{NA}'} \qquad [1.12]$$

with NA' being the image numerical aperture.

1.2.3. *Image formation under coherent light*

We now know the response of the microscope to a point excitation, let us examine the formation of images for a more complex object. Consider an extended object placed in the object plane and illuminated by a monochromatic plane wave of amplitude $\mathcal{A}_0$. The object is characterized by its transmission coefficient $\tau(X,Y)$, which typically takes values whose modulus is between 0 and 1 but whose phase can be nonzero. The wave amplitude just after the object is denoted by $\mathcal{O}(X,Y) = \mathcal{A}_0\,\tau(X,Y)$: the object is a kind of amplitude and phase modulator of the incident plane wave.

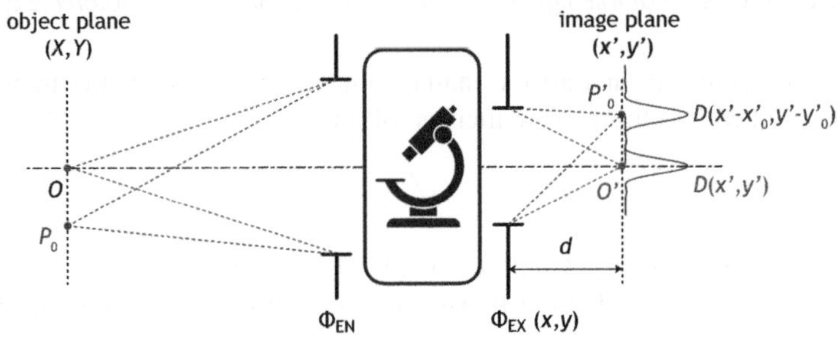

Figure 1.7. *Image formation in coherent light: image of an on-axis point source and an off-axis point source. For a color version of this figure, see www.iste.co.uk/sibarita/photonic.zip*

NOTATIONS.– O and O' are the respective centers of the object plane and the image plane, P_0 and P_0' are the two conjugate points, respectively, included in the object plane and the image plane, Φ_{EN} and Φ_{EX} are, respectively, the entrance and exit pupils, *d* is the distance between the exit pupil and the image plane, we denote $P'(x',y')$ any point on the image plane, and $M(x,y)$ any point on the exit pupil plane.

Let $\mathcal{D}(x',y')$ be the microscope percussion response. The conjugation is said to be aplanatic if and only if the percussion response is the same at all points of the image plane. In concrete terms, this means that field aberrations are zero. More reasonably, the term isoplanatic field is used to designate the area of the field for which the percussion response is rigorously identical to that of the point O. Consider a point source object placed at the point P_0 of coordinates (x_0, y_0) in the isoplanatic field. The spatial distribution of the complex amplitude in the image plane is centered at point P_0' geometric image of P_0 of coordinates $(x_0' = \gamma X_0, y_0' = \gamma Y_0)$ with γ being the transverse magnification, which has a form identical to the percussion response

$\mathcal{D}(x', y')$. It is precisely the condition of aplanatism that requires that the shape of the percussion response is not modified in the field:

$$\delta(X, Y) \xrightarrow{\text{microscope}} \mathcal{D}(x', y') \qquad \text{object point at } O$$

$$\delta(X - X_0, Y - Y_0) \xrightarrow{\text{microscope}} \mathcal{D}(x' - x'_0, y' - y'_0) \qquad \text{object point at } P_0$$

Let P' be a point with coordinates (x', y') in the image plane. The infinitesimal contribution of the point P_0 to the overall amplitude of the image is evaluated by considering a small surface element with an area $\mathrm{d}X_0\mathrm{d}Y_0$ centered at P_0 and which can be regarded as a point source. This contribution is written as:

$$\mathrm{d}\mathcal{A}(x', y') = \mathcal{O}(X_0, Y_0) \times \mathcal{D}(x' - x'_0, y' - y'_0)\,\mathrm{d}X_0\mathrm{d}Y_0$$

Using the image plane coordinate system only, we get:

$$\mathrm{d}\mathcal{A}(x', y') = \mathcal{O}(\frac{x'_0}{\gamma}, \frac{y'_0}{\gamma}) \times \mathcal{D}(x' - x'_0, y' - y'_0)\,\frac{\mathrm{d}x'_0}{\gamma}\frac{\mathrm{d}y'_0}{\gamma}$$

It is usual to omit the proportionality term in $1/\gamma^2$ which is relevant only in photometric terms. In addition, we recognize in the first term the expansion of the object amplitude. It is the perfect image in the sense of geometric optics, that is, the amplitude of the object magnified by a factor γ corresponding to the transverse magnification, we denote it by $\mathcal{O}'(x'_0, y'_0)$:

$$\mathrm{d}\mathcal{A}(x', y') \propto \mathcal{O}'(x'_0, y'_0) \times \mathcal{D}(x' - x'_0, y' - y'_0)\,\mathrm{d}x'_0\mathrm{d}y'_0$$

Since lighting is coherent, the overall amplitude at point $P'(x', y')$ is obtained by summing the amplitudes for each point P_0 of the object:

$$\mathcal{A}(x', y') \propto \iint \mathcal{O}'(x'_0, y'_0) \times \mathcal{D}(x' - x'_0, y' - y'_0)\,\mathrm{d}x'_0\mathrm{d}y'_0$$

This is the convolution of the geometric image $\mathcal{O}'$ with the percussion response $\mathcal{D}$:

$$\mathcal{A}(x', y') \propto \mathcal{O}'(x', y') * \mathcal{D}(x', y') \qquad [1.13]$$

The image irradiance is given by the square modulus of the amplitude:

$$\mathcal{I}(x', y') = \mathcal{A}(x', y')\mathcal{A}^*(x', y') \propto |\mathcal{O}'(x', y') * \mathcal{D}(x', y')|^2 \qquad [1.14]$$

More specifically, the convolution of the geometric image with the percussion response in amplitude will lead to losing details: it is as if the complex amplitude of the image were drawn by tracing the geometric image with a brush too thick (the amplitude percussion response).

1.2.4. *Microscope resolution (diffraction-limited)*

The aperture diaphragm of a microscope is usually placed in the back focal plane of the objective and has a circular shape. When there are no aberrations, the image size of a dot is inversely proportional to the size of the diffracting aperture: diffraction limits the resolution. Two point objects can be separated if their Airy disks are not too overlapping. The Rayleigh criterion is generally used: when the top of one of the diffraction patterns corresponds to the first zero minimum of the other, the peaks are too close for an observer to distinguish them (see Figure 1.8).

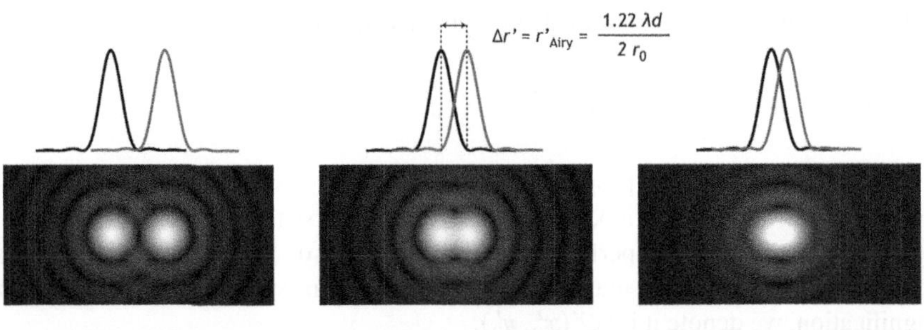

Figure 1.8. *Rayleigh criterion. Diffraction of the image of two incoherent object points by a microscope, under the conditions of the paraxial approximation. For a color version of this figure, see www.iste.co.uk/sibarita/photonic.zip*

Resolution designates the minimal distance between two incoherent object points of the same intensity necessary to differentiate them with the optical instrument under consideration. In the example where the exit pupil of the optical system is circular with a radius of r_0, each of these object points forms an Airy pattern with radius $r'_{\text{Airy}} = 0.61\,\lambda d/r_0$ in the image plane (equation [1.12]). The images can therefore be distinguished in the sense of the Rayleigh criterion if they are separated by a distance $\Delta r'$ such that (see Figure 1.8):

$$\Delta r' = r'_{\text{Airy}} = 0.61\,\frac{\lambda d}{r_0} = 1.22\,\frac{\lambda d}{2\,r_0} = 1.22\,\frac{\lambda}{2\,\alpha'} = 1.22\,\frac{\lambda}{2\,\text{NA}'} \qquad [1.15]$$

When it is desirable to improve the resolution of a microscope (i.e. decrease $\Delta r'$), one way to do this is to work at the lowest possible wavelengths. This is true for all far-field imaging techniques. Therefore, it is also true for electron microscopy techniques that use wavelengths associated with the electron beam that are about 10^4 times shorter than visible wavelengths. The resolution can also be improved by increasing the numerical aperture of the lens through the use of an immersion liquid. On the contrary, this implies keeping an eye on the correction of optical aberrations, which potentially becomes increasingly critical for large apertures, especially when involving possibly thick biological samples, unless a correction system (possibly dynamic) such as adaptive optics, for example, is implemented.

1.2.5. *Abbe's theory of image formation*

Another way of considering the effect of diffraction on image formation is to reason, not in direct space as previously, but in reciprocal space, i.e. in Fourier space or space of spatial frequencies. This theory of image formation in a microscope was formulated by Ernst Abbe at the end of the 19th century. Abbe treats the object as a diffraction grating that diffracts light. Diffracted rays therefore combine with the undeviated zeroth-order ray, on either side and according to the successive diffraction orders: -1 and +1, -2 and +2, etc. All of these orders contain some information about the object under study. The aperture of the objective lens limits the number of diffraction orders likely to travel through the system: given that structures strongly diffract light, namely corresponding to a very fine level of detail and as such to very high spatial frequencies, they will produce diffraction orders with high angles that will not be collected by the lens. Since they are nonexisting in image space, the formation of an image that is a faithful reconstruction of the object will not be possible.

1.2.5.1. *How lighting affects image resolution*

The schematic diagram is shown in Figures 1.9 to 1.13. To simplify, the objective $\mathcal{L}$ will be treated as a thin lens and works in $4f$ conjugation with a magnification γ of -1. The aperture diaphragm of the system is placed in the back focal plane of the objective lens $\mathcal{L}$. This is therefore the exit pupil of the system. Its conjugate in object space is projected in the object focal plane of the condenser $\mathcal{C}$: this is the entrance pupil. The assembly is assumed to be free of aberrations and therefore limited by diffraction. The lighting is coherent when the source, placed in the object focal plane of the condenser (thus in the plane of the entrance pupil), is point-like (see Figure 1.9). It is partially coherent when it is spatially extended in the plane of the entrance pupil, while remaining smaller than the entrance pupil (see Figures 1.10 and 1.11). It is completely incoherent when the source completely covers the entrance pupil (see Figures 1.12 and 1.13). The object is placed in the object focal plane of the objective lens $\mathcal{L}$. In coherent lighting, this object is illuminated by a monochromatic plane wave parallel to the optical axis. In incoherent lighting (partially or completely), the object is illuminated by an infinity of monochromatic plane waves, more or less inclined with respect to the optical axis depending on the initial point of the source from which they originate.

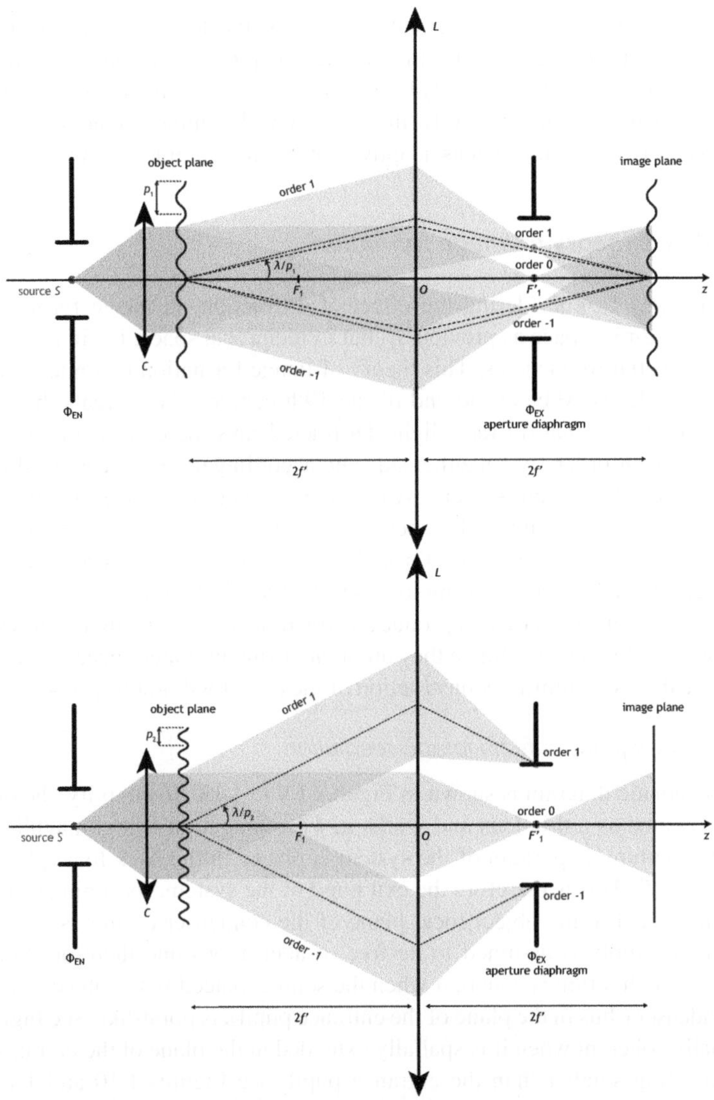

Figure 1.9. *Transmission of modulation in coherent lighting. In the Fourier plane, combined with the exit pupil, we observe a Fraunhofer-type diffraction pattern. In the case of a single-frequency pattern, as long as order 1 and order -1 are sufficiently close to the optical axis (top diagram: low spatial frequencies) to travel through the exit pupil, the pattern is reconstructed with a modulation 1. On the contrary, beyond the cut-off frequency* $w'_C = R_{EX}/\lambda f'$ *(bottom diagram) only the zero frequency is transmitted, the modulation of the pattern image drops sharply to 0. For a color version of this figure, see www.iste.co.uk/sibarita/photonic.zip*

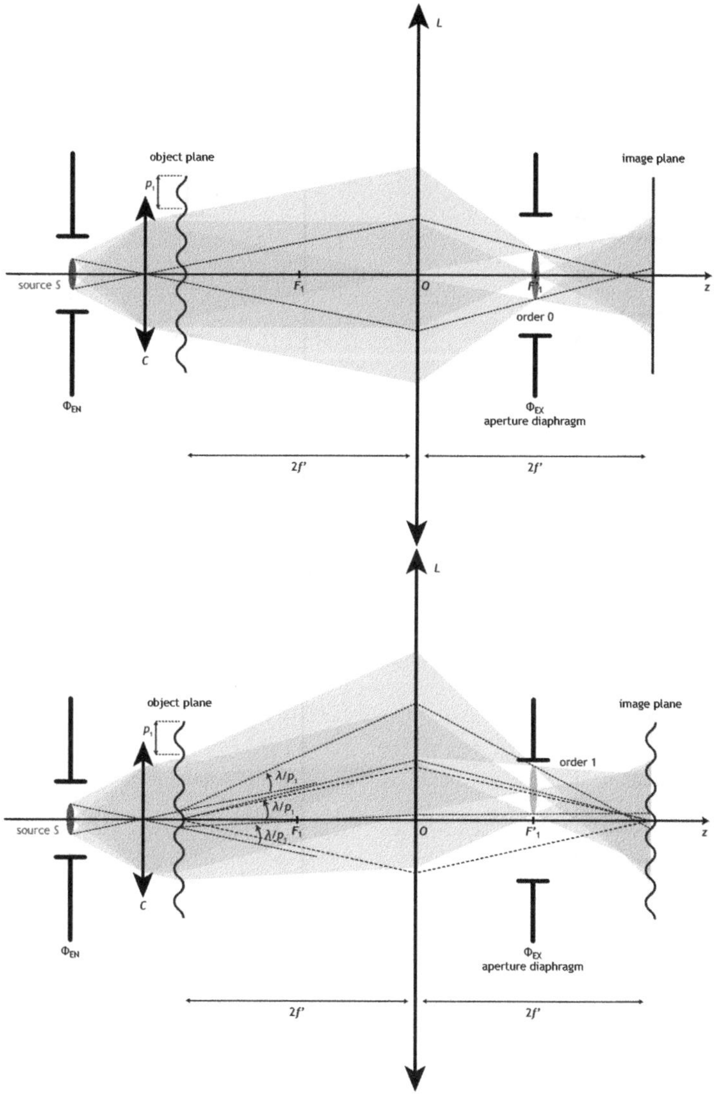

Figure 1.10. *Transmission of modulation in partially incoherent lighting: the image of the source is smaller than the exit pupil. The zero frequency is therefore transmitted with a modulation of 1 (top diagram). In addition, in the case of low spatial frequencies, if the source image is small enough that its first order is also entirely included in the exit pupil (bottom diagram), all of the light diffracted by the target travels through the optical system: the low spatial frequencies are also transmitted with a modulation of 1. For a color version of this figure, see www.iste.co.uk/sibarita/photonic.zip*

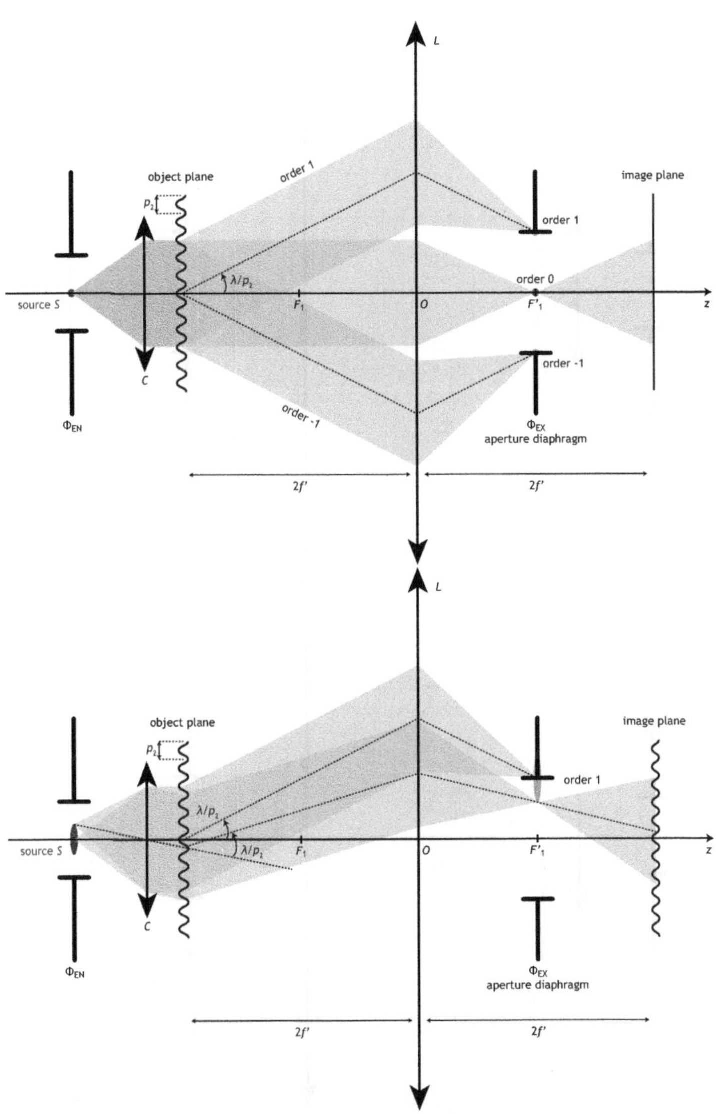

Figure 1.11. *Transmission of modulation in partially incoherent lighting: the cut-off frequency is greater than in coherent lighting. For high spatial frequencies (bottom diagram), as the source is spatially extended, the first order of its image can be shifted further in the plane of the exit pupil without all of the light being blocked. It is therefore possible to reconstruct the image of the pattern, even if the modulation will no longer be equal to 1. The exact value of the cut-off frequency depends on the size of the source. For a color version of this figure, see www.iste.co.uk/sibarita/photonic.zip*

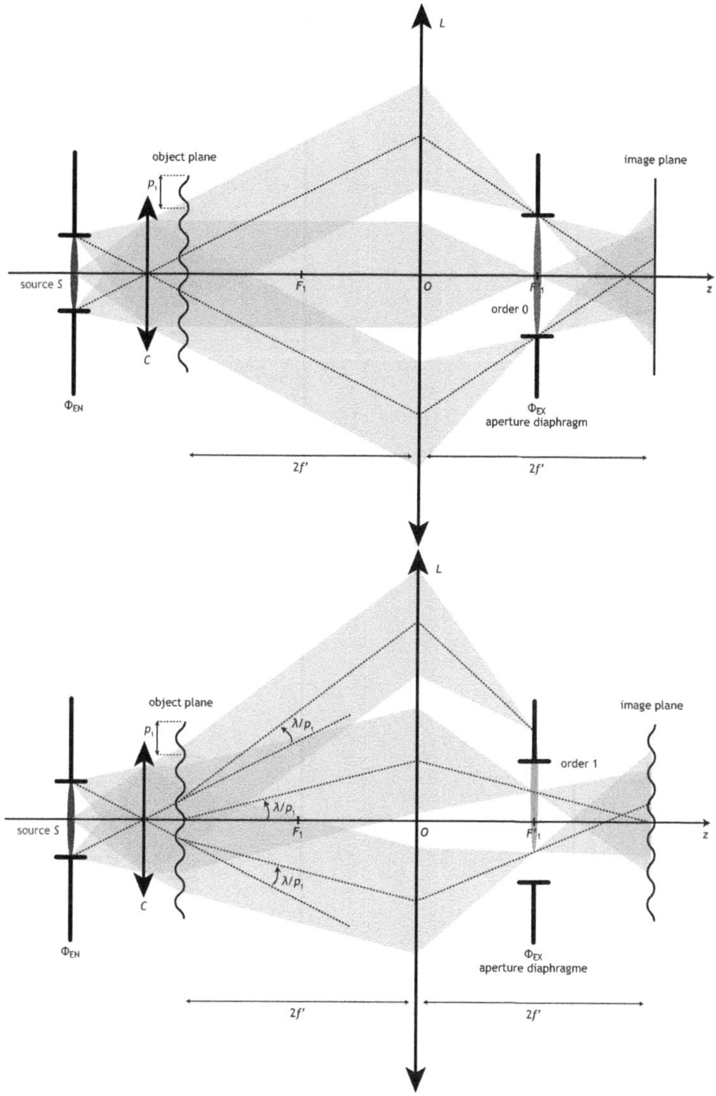

Figure 1.12. *Transmission of modulation in perfectly incoherent lighting: the zeroth-order of the source image completely covers the system exit pupil. The top diagram highlights the fact that the zero frequency is perfectly transmitted regardless of the object. On the contrary, due to the size of the source image in the exit pupil, the diagram at the bottom shows that even the low frequencies will never be perfectly transmitted: the first order of the source image never travels completely through the exit pupil. For a color version of this figure, see www.iste.co.uk/sibarita/photonic.zip*

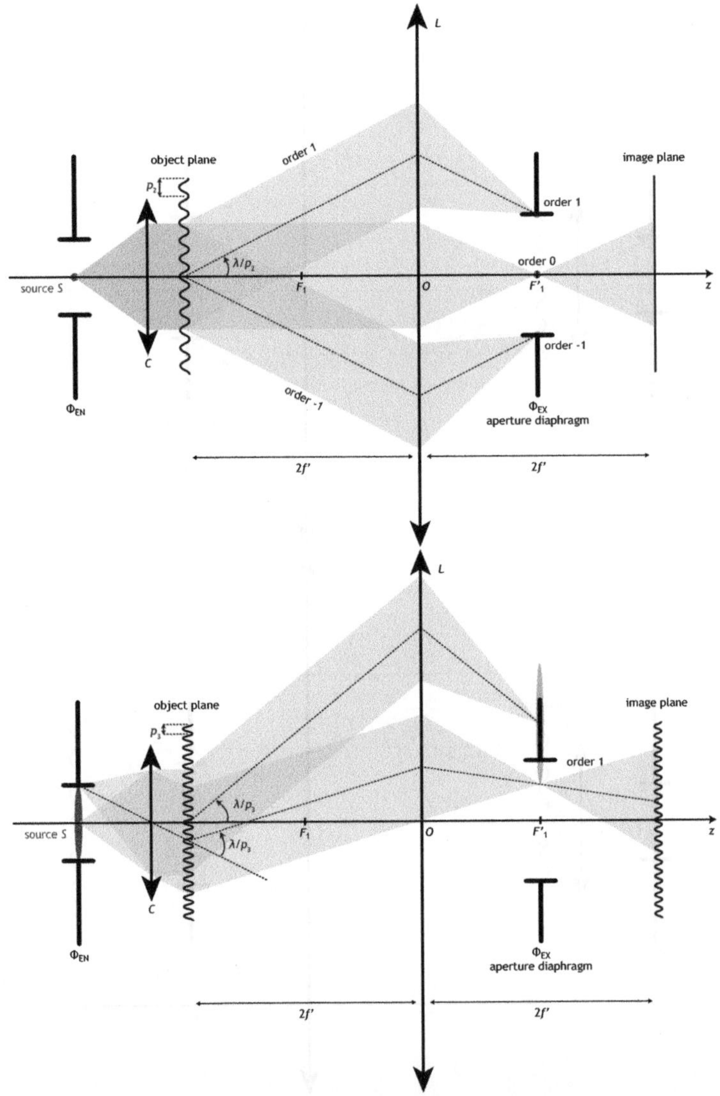

Figure 1.13. *Transmission of modulation in perfectly incoherent lighting: the cut-off frequency is twice as large as in coherent lighting. In incoherent lighting, since the source image completely covers the exit pupil, its first order can be shifted by one pupil size before being completely blocked. The corresponding spatial frequency* $w'_I = 2R_{EX}/\lambda f'$ *in the bottom diagram is indeed twice as greater than in the top diagram* $w'_C = R_{EX}/\lambda f'$*. For a color version of this figure, see www.iste.co.uk/sibarita/photonic.zip*

Consider as an object a target alternately opaque and transparent to light in the direction (Oy), of step p_i. The pattern of step p_1 represents objects with a low spatial frequency, the pattern of step $p_2 < p_1$ represents objects with a high spatial frequency, and the pattern with $p_3 < p_2$ represents objects with a very high spatial frequency. Each of these patterns therefore has three spatial frequencies: 0, $1/p_i$ and $-1/p_i$. These are projected as light source points in the image focal plane (the Fourier plane) at coordinates $(0, 0)$, $(0, \lambda f'/p_i)$ and $(0, -\lambda f'/p_i)$. The target actually behaves like a diffraction grating that produces three orders: order 0 is carrying the zero frequency that propagates in the direction of geometric optics, order 1 and order -1 are carrying the modulation to frequencies $\pm 1/p_i$ that are diffracted in directions $\pm\lambda/p_i$ (see Figure 1.9). As long as the 1st order and the -1st order are sufficiently close to the optical axis to go through the exit pupil, the pattern is reconstructed with a modulation 1. On the contrary, beyond the cut-off frequency $w'_C = R_{\text{EX}}/\lambda f'$, only the zero frequency is transmitted, the modulation of the pattern image drops sharply to 0.

In incoherent lighting (partially or completely), the pattern also behaves like a diffraction grating that produces three orders: order 0 (zero frequency, direction of the geometric optics), orders +1 and -1 (frequencies $\pm 1/p_i$, directions $\pm\lambda/p_i$). This effect occurs for each of the plane waves illuminating the object, whether they are parallel to the optical axis or slanted with respect thereto. Therefore, in the Fourier plane, the previous diffraction pattern of the coherent lighting is convolved with the source image in the plane of the exit pupil.

For partially incoherent lighting, this will have two main consequences:

1) For low spatial frequencies, if the image of the source is small enough that orders ± 1 are also fully included in the exit pupil (see Figure 1.10), all of the light diffracted by the target travels through the optical system: the low spatial frequencies are transmitted with a modulation of 1.

2) For high spatial frequencies, since the source is spatially extended, the orders ± 1 of its image can be shifted further in the plane of the exit pupil without all of the light being hidden. It is therefore possible to reconstruct the image of the pattern, even though the modulation will no longer be equal to 1. The cut-off frequency is greater than in coherent lighting (see Figure 1.11).

In the case of completely incoherent lighting, the source completely covers the system entrance pupil. Its image therefore completely covers the exit pupil. The convolution of the source image with the diffraction pattern under coherent illumination will produce two slightly different consequences:

1) Only the zero frequency is perfectly transmitted independently of the object. Indeed, due to the size of the image of the source in the exit pupil, the orders ± 1 of the source image never travel completely through the exit pupil: part of the light needed to reconstruct the target with a modulation 1 is systematically lost. Therefore, the target is never reconstructed with a modulation of 1 (see Figure 1.12).

2) The cut-off frequency is twice as high as in coherent lighting. In incoherent lighting, since the image of the source completely covers the exit pupil, its first order can be shifted by one pupil size before being completely blocked. The corresponding spatial frequency $w'_I = 2R_{\text{EX}}/\lambda f'$ in the bottom diagram is indeed twice as large as in the top diagram $w'_C = R_{\text{EX}}/\lambda f'$ (see Figure 1.13).

1.2.5.2. *Notion of modulation transfer function*

The *modulation transfer function* (or MTF) is a function that characterizes the ability of the optical system to reproduce contrast according to the level of details of the object: in other words, its ability to transmit the spatial frequencies of the object. Based on the intuitive considerations developed in the previous section, we can now trace the shape of the MTFs in the three lighting conditions. In coherent lighting, the MTF_C is binary with an abrupt transition to the cut-off frequency $w'_C = R_{\text{EX}}/\lambda f' = \text{NA}'/\lambda$, with NA' being the image numerical aperture of the system. In partially incoherent lighting, the MTF_{PI} is 1 for low frequencies and then gradually decreases until it reaches a value of 0 for a cut-off frequency greater than w'_C. In completely incoherent lighting, the MTF_I is strictly decreasing from the zero frequency and reaches a value of 0 for a cut-off frequency $w'_I = 2\,w'_C = 2\,R_{\text{EX}}/\lambda f' = 2\,\text{NA}'/\lambda$. These results are summarized in Figure 1.14.

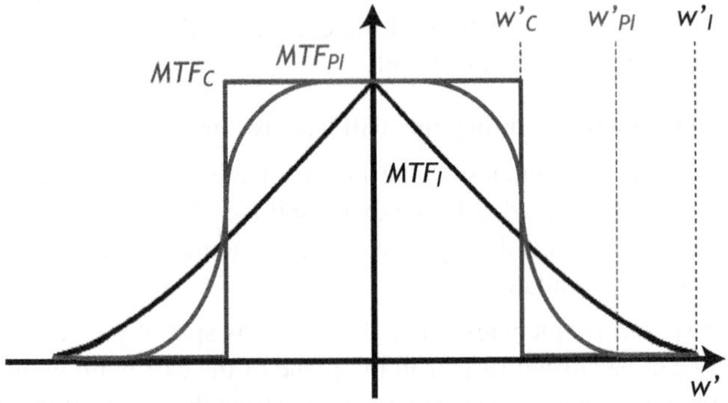

Figure 1.14. *Shape of the modulation transfer function in the different light modes. In blue for coherent light, in red for partially incoherent light, in black for completely incoherent light. For a color version of this figure, see www.iste.co.uk/sibarita/photonic.zip*

By reasoning in Fourier space, it is also possible to demonstrate that for a diffraction-limited optical system, the MTF_C is in fact equal to the normalized Fourier transform of the coherent percussion response of the microscope: in other words, it is equal to the normalized pupil function. In contrast, the MTF_I is equal to the normalized Fourier transform of the PSF (which is also the incoherent percussion response): in other words, the autocorrelation of the normalized pupil function. In addition, in a microscope or in any real optical system, most geometric and chromatic aberrations, in addition to manufacturing or alignment defects, decrease the value of the ideal MTF. Conversely, the PSF is equally affected and then typically possesses a larger size than the Airy disk, or even a shape without symmetry of revolution, which may even differ depending on the field area under consideration.

1.3. Optical sectioning techniques in fluorescence microscopy

For the purposes of imaging a biological sample in fluorescence light microscopy, the first problem that emerges is the fact that the samples are very often thicker (a few micrometers) than the depth of field of the microscope objective ($\approx$ 600 nm). The fluorescence produced by the out-of-focus planes of the sample therefore appears blurry on the detector, which considerably degrades the image quality. Technological innovations in fluorescence microscopy therefore include extracting the optical section corresponding to the plane of focus, namely eliminating the out-of-focus fluorescence. An example of a U2OS[3] cell labeled with *Alexa Fluor 532* is presented as an illustration of this principle in Figure 1.15.

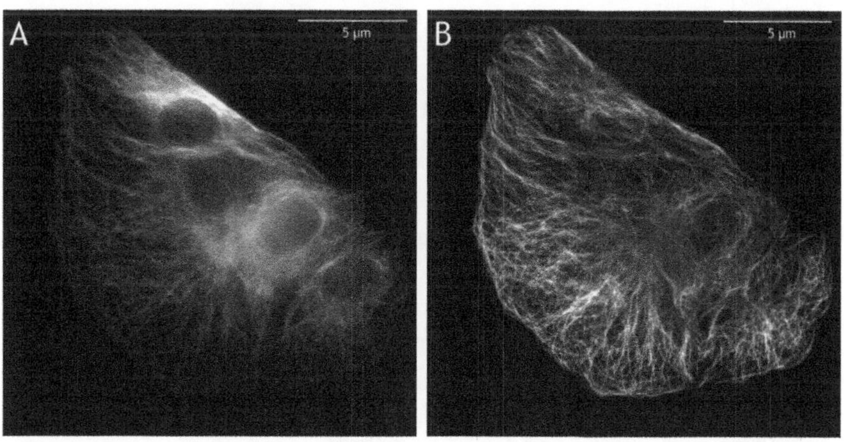

Figure 1.15. *A. Comparison between two images of the same sample without and with optical sectioning. A. Epifluorescence image of a U2OS cell with alpha tubulin labeled with* Alexa Fluor 532. *B. TIRF image of the same cell (Images © Laurent Le)*

3. The U2OS cell line, originally known as the 2T line, is a commercial cell line widely used in biomedical research. It was initially cultivated from the bone tissue of a 15-year-old human female suffering from osteosarcoma.

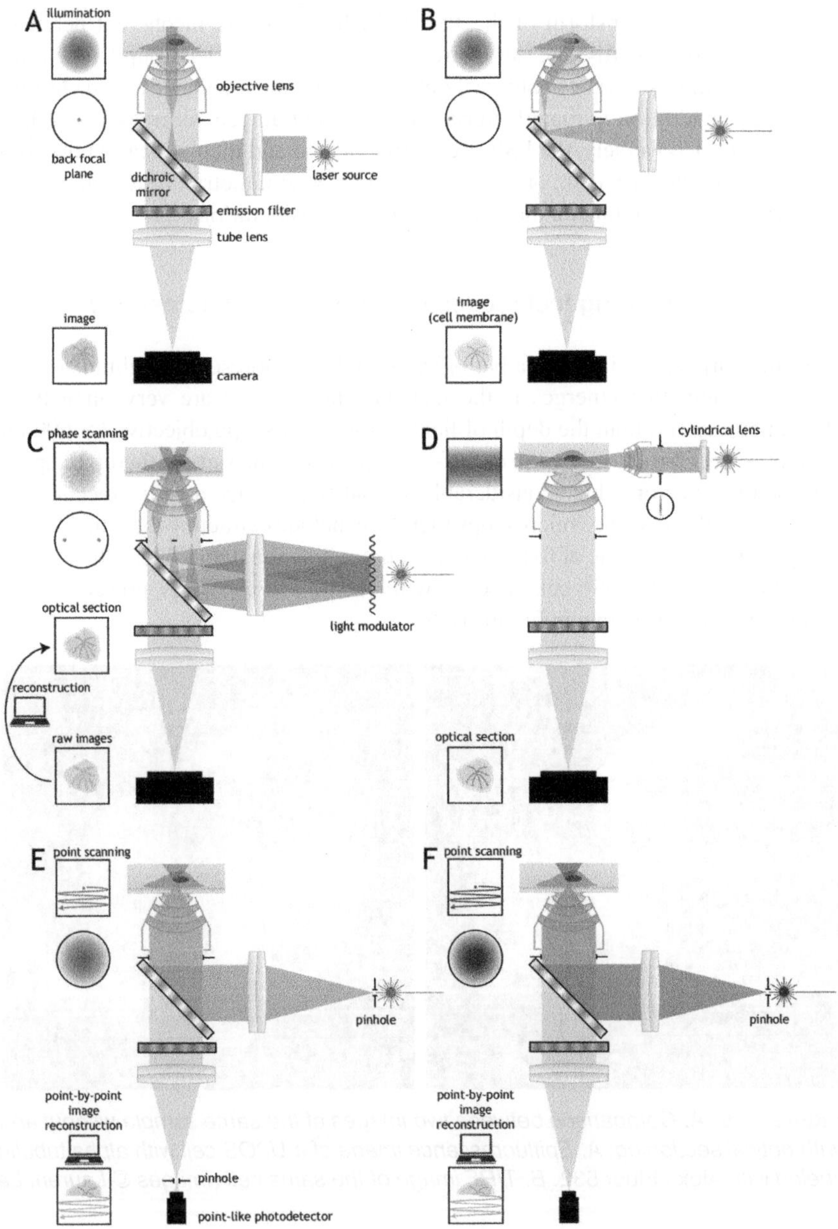

Figure 1.16. *Optical sectioning techniques. A. Conventional epifluorescence configuration; B. TIRF configuration; C. structured illumination; D. light sheet illumination; E. confocal configuration; F. two-photon microscopy (diagrams inspired by Schermelleh et al. (2019)). For a color version of this figure, see www.iste.co.uk/sibarita/photonic.zip*

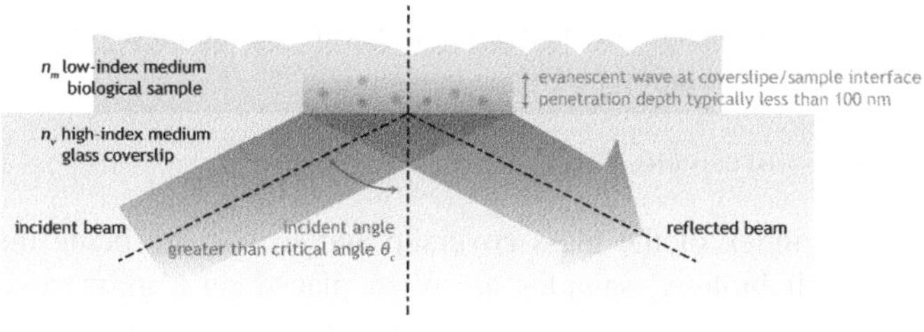

Figure 1.17. *Total internal reflection microscopy. Principle of TIRF excitation: an excitation beam incident on an interface with an incidence angle greater than the critical angle θ_c is fully reflected. This total internal reflection generates an evanescent wave on the other side of the interface whose penetration depth is typically of about a hundred nanometers. For a color version of this figure, see www.iste.co.uk/sibarita/photonic.zip*

Out-of-focus fluorescence filtering is not simple because fluorescence is an incoherent mechanism: it is therefore not possible to perform filtering in the Fourier plane as can be done in coherent imaging. On the contrary, fluorescence excitation is generally carried out using laser sources, therefore through a coherent mechanism. In the vast majority of optical sectioning techniques, it is therefore at the excitation stage that filtering occurs. These different approaches are schematized in Figure 1.16: some are possible in wide field, whereas others require scan-based point-by-point acquisition.

1.3.1. *Wide-field techniques*

1.3.1.1. *Evanescent-wave optical sectioning*

The use of evanescent waves for imaging biological objects below the diffraction limit is not unique to optical near-field techniques. For example, Total Internal Reflection Fluorescence (TIRF) microscopy uses an evanescent wave excitation mode: the fluorescence is confined to a very thin optical section of the sample, immediately adjacent to the glass coverslip. Conversely, Supercritical Angle Fluorescence (SAF) microscopy consists of selectively detecting the fraction of the near-field emission of fluorophores that has become propagating.

TIRF MICROSCOPY.– Wide-field optical sectioning in total internal reflection fluorescence microscopy was proposed by Daniel Axelrod in 1981. The principle consists of taking advantage of the evanescent wave generated within the sample when the angle of the illumination beam is greater than the critical angle θ_c (see Figure 1.17). This corresponds to the angle beyond which the incident light on the interface is completely reflected: this phenomenon is known as *total internal reflection*. The value of the critical angle depends on the refractive index of the media on either side of the interface and is given by relation [1.16]:

$$\theta_c(n_m, n_g) = \arcsin(n_m/n_g) \quad [1.16]$$

with n_v being the index of the glass coverslip and $n_m < n_v$ being the index of the medium. In cell biology, samples are often placed on a glass coverslip with index $n_g \approx 1.51$ and immersed in an aqueous medium with index $n_m \approx 1.33$, which leads to a critical angle $\theta_c \approx 62°$. Therefore, for any angle of incident light $\theta \geq \theta_c$, the excitation beam is completely reflected, and an evanescent non-propagative electromagnetic field is created at the interface. The irradiance relative to this evanescent field $I(z)$ decreases exponentially with the axial distance from the interface z following the relation:

$$I(z) = I_0 \exp\left(\frac{-z}{\Delta_z(\theta)}\right) \quad \text{with} \quad \Delta_z(\theta) = \frac{\lambda_{\text{exc}}}{4\pi} \times \frac{1}{\sqrt{\left(n_v{}^2 \sin^2(\theta) - n_m{}^2\right)}} \quad [1.17]$$

with I_0 being the irradiance at the interface and $\Delta_z(\theta)$ being the depth of penetration of the evanescent field at the excitation wavelength λ_{exc}. As an example, for an angle of TIRF incidence $\theta = 66.5°$ and at the wavelength $\lambda_{\text{exc}} = 488$ nm, the penetration depth will be $\Delta_z \approx 100$ nm within a biological sample. Under these conditions, only fluorophores located in the first 100 nanometers after the interface will be excited.

SAF MICROSCOPY.– The illumination of a glass/water interface at an angle of incidence greater than the critical angle θ_c generates an evanescent wave over a hundred nanometers on the surface of the same interface. According to the principle of reversibility of light propagation, the inverse of this phenomenon also holds: an evanescent wave located near an interface can be coupled and propagates on the other side of the interface beyond the critical angle. This phenomenon of supercritical emission is often referred to for fluorescence by its acronym SAF or *forbidden light*. Fluorophores behave like dipole nano-antennas whose emission includes near-field and far-field components. While the latter is associated with the mode of light propagation, the near field is associated with the evanescent mode located around each fluorophore. Analogous to the evanescent excitation wave associated with the TIRF configuration, the intensity associated with the near-field emission of a fluorophore decreases exponentially further away from the latter until it becomes negligible beyond a distance equivalent to the emission wavelength. Provided that the optical index of the medium surrounding the n_m fluorophore is lower than that of the glass n_g (configuration of the indexes identical to the TIRF configuration), this evanescent field can become propagative. Consequently, as shown in Figure 1.18, the fluorescence emission propagating in the glass under these conditions consists of two components: the far-field fluorescence that is emitted at angles $\theta < \theta_c$ (UAF) and the fluorescence associated with the part of the near-field that has become propagative that is emitted at angles $\theta \geq \theta_c$ (SAF). The SAF contribution is far from negligible since it can represent 34% of the total emission and 50% of the fluorescent emission that propagates inside the glass coverslip, as described in Ruckstuhl et al. (2000) and Fort and Grésillon (2008). Moreover, if the intensity of the UAF far-field component is independent of the distance z of the fluorophore to the interface, the SAF intensity exponentially decreases with z.

The distinction of the SAF and UAF components of a fluorophore can be achieved in particular in the back focal plane of an aplanatic objective (or in a conjugate plane therewith) (Barroca et al. 2011). Due to aplanatism, the relationship of Abbe's sines is verified, and the fluorescence emission of a given fluorophore in the direction θ is projected into the back focal plane – where the aperture diaphragm of the objective is also located – with the shape of a circle of radius:

$$r(\theta) = f' \, n_i \sin\theta \qquad [1.18]$$

where n_i is the index of the immersion oil (equal to the index of the glass) and f' is the focal length of the lens. In the back focal plane, the UAF component thus appears as a disk with radius $r_{\text{UAF}} = f' \, n_i \sin\theta_c$, while the SAF component, when it exists, appears as a ring with an inner radius r_{UAF} and an outer radius $r_{\text{SAF}} = f' \, \text{NA}'$ (see Figure 1.18).

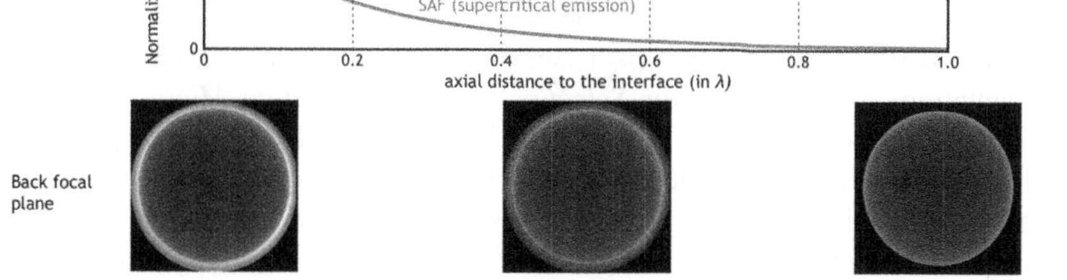

Figure 1.18. *Principle of supercritical emission. The SAF emission exponentially decreases with the distance between the fluorophore and the interface and is projected in the shape of a ring onto the back focal plane (BFP) of the objective. For a color version of this figure, see www.iste.co.uk/sibarita/photonic.zip*

1.3.1.2. *Structured illumination*

The principle of structured illumination is to modulate the excitation in the plane of focus by projecting a periodic pattern. This pattern has a specific spatial frequency. Therefore, it will only be visible over a certain thickness within the sample, which is actually the value of the depth of field at the corresponding frequency. In other words, images acquired with structured illumination possess two components: the fluorescence originating from the focus plane is modulated by the illumination pattern, while the fluorescence from the out-of-focus planes is unmodulated.

The modulation of the illumination in the focus plane can be achieved by optical conjugation of a periodic physical grid (such as a Ronchi pattern) or through the use of a matrix of micro-mirrors. Another approach consists of achieving sinusoidal illumination by creating an interference pattern with the $+1$ and -1 diffraction orders of a grating (for instance, using a spatial light modulator) or still by means of a laser split into two beams that will directly interfere within the sample. The reconstruction of the optical cross-section can be done by combining three or four images, respectively, shifted by a third or a quarter of a period (Neil et al. 1997; Karadaglić and Wilson 2008) or by associating a modulated image and a uniform image (Lim et al. 2008, 2011).

In addition to optical sectioning, structured illumination techniques can improve the spatial resolution of images by a factor of 2 compared to the diffraction limit, as described in Gustafsson (2000) and Gustafsson et al. (2008). This is because the high spatial frequencies of the sample merge with the spatial frequency of the illumination pattern, which produces a Moiré pattern at the lower difference frequency. If the latter is not cut off by the microscope's MTF, some details that were initially excluded can be extracted from the modulated image with digital processing. These approaches are the subject of Chapter 7 of this book.

1.3.1.3. *Light-sheet illumination*

Light-sheet illumination is based on the shaping of the excitation beam on a path detached from the detection path (see Huisken and Stainier 2009). The usual implementation, called SPIM (for *Selective Plane Illumination Microscopy*), uses a cylindrical lens that focus the excitation laser beam following a single lateral direction perpendicular to the detection: the sample is illuminated transversally by a sheet of light whose typical thickness lies in the micrometer range. The axial position of the excitation beam is controlled by a scanning system so as to allow recording images in 3D, while minimizing illumination in unobserved axial planes. The orthogonal detection to the excitation beam is in fact a standard epifluorescence detection combining a matrix detector and a spectral filtering system of the emission.

This strategy allows for the rapid acquisition of wide fields of view and optical sectioning over a wide axial range. Indeed, in-depth imaging is limited only by the absorption and scattering properties of the sample. On the contrary, implementations remain complex and require adapting the techniques for the preparation of the samples and associated media. In addition, in order to take full advantage of this technique in terms of 3D imaging of large-sized samples of small animals (drosophila or zebrafish, for example), light-sheet microscopes are increasingly integrating wavefront correction strategies for preserving the quality of optical sectioning at depth as described, for example, in Hubert et al. (2019). For a more in-depth presentation of the instrumental principles of light-sheet microscopy, see section 6.4 in Chapter 6 of this book.

1.3.2. *Point-scanning techniques*

The invention of confocal photon microscopy methods dates back to the mid-1950s. At that time, Marvin Minsky was faced with the need to observe neuronal cells in thick sections of nerve tissue. In 1957, he developed a prototype microscope with a scanning stage and equipped with what was then called a double focusing system. This device was quickly named the confocal system. This technique, however, remained unused for a long time, as it required the use of a point source delivering a high luminous intensity and the image had to be formed on a high-remanence screen. In fact, it was not until the early 1980s that the production of low-cost laser sources and the exponential development of computer science allowed this microscopy technique to find the place it now occupies in the field of biology.

1.3.2.1. *Scanning confocal microscopy*

The main technique for point-scanning optical sectioning is confocal scanning fluorescence microscopy. It is based on the use of a focused laser beam for excitation and the positioning of a *pinhole* in front of the detector. Therefore, at a given moment, only a tiny part of the specimen is strongly illuminated and will emit fluorescence. The photons emitted by the fluorophores are detected in a point-like manner with a photodetector during sample scanning. The pinhole placed in front of the photodetector and positioned in a conjugated plane with the objective focus plane rejects stray light from out-of-focus planes. The term *confocal* originates from the matching of three points: the point light source (usually a laser source), the illuminated point located at the object focus of the objective and the pinhole located in the final image plane of the device. In addition to following the conditions of classical optics (choice of wavelength, objective numerical aperture, limitation of aberrations, etc.), the adjustable diameter of the pinhole largely determines the thickness of the optical section obtained after reconstruction. The optimal spatial resolution is obtained when it is adjusted to the diameter of the Airy disk.

As in any optical instrument, the confocal microscope is governed by the limitations imposed by diffraction. Nonetheless, the confocal configuration allows for a gain in lateral and axial resolution compared to conventional fluorescence microscopy, as shown in Figure 1.19. To understand how this improvement occurs, consider a confocal beam scanning a fluorescent nanosphere whose size is smaller than the diffraction limit of the excitation beam.

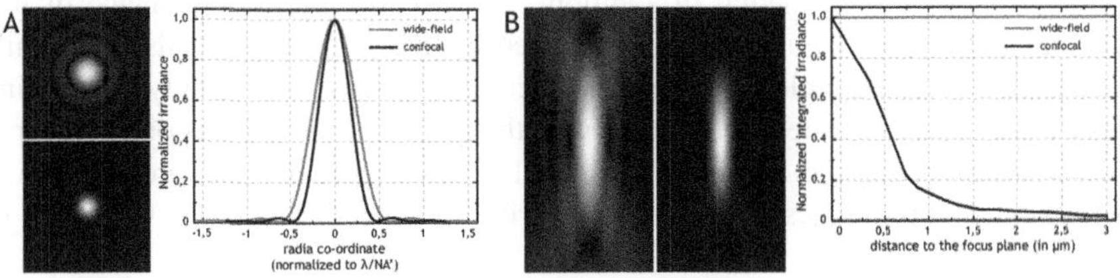

Figure 1.19. *Lateral and axial resolutions in wide-field and confocal configurations. A. Lateral PSF in the focus plane in wide-field (top) and confocal (bottom) configuration. The corresponding normalized irradiances are, respectively, plotted in red and green. B. Axial PSF in a plane containing the optical axis in the wide-field (left) and confocal (right) configuration. Integrated normalized irradiances for different out-of-focus values with respect to the focus plane are plotted in red and green, respectively. For a color version of this figure, see www.iste.co.uk/sibarita/photonic.zip*

The beam itself projects an Airy pattern on the plane of the specimen, which, in the plane of focus, appears as a bright central disk surrounded by concentric rings of progressively fainter irradiances. When this pattern gets closer to the nanosphere, the first interaction occurs between a portion of the outer rings and the nanosphere. This will give rise to very low excitation of the latter, the image of which (also an Airy pattern, but at the emission wavelength) will be projected onto the detection pinhole. The latter is always aligned with the center of the illumination Airy spot, so only a portion of the outer ring of the nanosphere image will propagate to the detector. Given that the nanosphere is little excited and only a small part of its emission is collected, only an extremely small part of the fluorescence emitted by the bead will be collected at this stage. As the illumination beam gets nearer to the nanosphere, the irradiance of the excitation increases and collection is now done from an area of the emission Airy pattern that contains more emitted light. Finally, when the illumination Airy disk coincides with the nanosphere, the excitation and collection of the emitted fluorescence are optimized. The result is that the combination of scanning and pinhole attenuates the emission Airy disk, so that more of the irradiance detected is related to the actual position of the bead and not to the peripheral rings, which are usually no longer detectable. In addition, the diameter of the central disk is also reduced for the same reason. In more technical terms, the PSF of the confocal microscope is the product of the excitation and detection PSF.

Confocal scanning microscopy is probably the reference technique for 3D imaging of fluorescence-labeled biological samples from the cell to the tissue. It exists in the overwhelming majority of biological imaging platforms and has found many clinical applications. However, it still has two major limitations. First, a point-by-point scan of the field has to be performed to reconstruct an image of the sample: this acquisition modality considerably limits the time resolution of the technique (≈ 1 μs/pixel). The irradiance at the focal point is about 20,000 times greater in the confocal configuration (≈ 2 kW/cm^2) than in the wide-field configuration (≈ 100 W/cm^2): the biological samples are likely to rapidly degrade due to photobleaching, phototoxicity mechanisms or simply due to local heating. Nevertheless, this level of irradiance is transient because the operation of a confocal microscope is based on scanning, which is not the case with a wide-field microscope. To compare more precisely the risks in terms of photodamage, it would be necessary to take into account the acquisition time of the camera for wide-field techniques (a few tens of milliseconds) and the scanning speed of the confocal (of about one second to acquire an optical section), and to proceed with reasoning in terms of exposure and not irradiance as described in Jemielita et al. (2013).

A solution to improve the time resolution in confocal microscopy consists of multiplexing the excitation and to include wide-field detection: this is the principle of multipoint confocal microscopes and *spinning disks* (see Tanaami et al. 2002). However, this improvement in time resolution generally comes at the cost of a decrease in spatial resolution and/or in the quality of optical sectioning. Confocal microscopy and more generally in-depth optical sectioning through spatial filtering are the subject of section 6.2 in Chapter 6 of this book.

1.3.2.2. *Two-photon microscopy*

Two-photon microscopy, described by Denk et al. (1990), is also a technique based on a scanning acquisition, but in this case, it is the two-photon nonlinear absorption that confines the effective size of the excitation PSF. Since the two-photon absorption cross-section of most fluorophores is very small, two-photon excitation requires a high instantaneous irradiance that is accessible through the development of ultra-short pulse laser sources (of a few hundred femtoseconds) in the combination with high numerical aperture objectives. In addition, in order to excite the usual fluorophores typically emitting in the green, these sources are centered in the near-infrared (≈ 700–1,000 nm): these wavelengths are about 10 times less absorbed and scattered by biological samples than the visible range, which enables better light penetration into the tissues over several hundred micrometers. On the contrary, this gain is partially offset by the increase in the excitation wavelength which leads to a loss of spatial resolution compared to confocal microscopy. Moreover, although reduced to the focal volume, the damage caused within the cellular medium under two-photon excitation is difficult to characterize for higher values of instantaneous local irradiance (a few TW/cm^2), according to De and Goswami (2009). The main limitation to the democratization of this technique remains, nonetheless, the high cost of femtosecond sources and the relative complexity of its implementation. Multi-photon microscopy is presented in further detail in section 6.3 of Chapter 6 of this book.

1.4. Conclusion

In this chapter, the foundation for the most commonly used light microscopy approaches for applications in biology was presented. The various limitations were highlighted, in particular in terms of spatial resolution, optical sectioning, acquisition speed and phototoxicity, and they will be discussed again in the following chapters, with different implementations that specifically address these limitations, thus paving the way for more dynamic observations that can reach the nanometric scale.

1.5. References

Axelrod, D. (1981). Cell-substrate contacts illuminated by total internal reflection fluorescence. *Journal of Cell Biology*, 89(1), 141–145.

Barroca, T., Balaa, K., Delahaye, J., Lévêque-Fort, S., Fort, E. (2011). Full-field supercritical angle fluorescence microscopy for live cell imaging. *Optics Letters*, 36(16), 3051–3053.

De, A.K. and Goswami, D. (2009). Exploring the nature of photo-damage in two-photon excitation by fluorescence intensity modulation. *Journal of Fluorescence*, 19(2), 381–386.

Denk, W., Strickler, J.H., Webb, W.W. (1990). Two-photon laser scanning fluorescence microscopy. *Science*, 248(4951), 73–76.

Fort, E. and Grésillon, S. (2008). Surface enhanced fluorescence. *Journal of Physics D: Applied Physics*, 41(1), 013001.

Gustafsson, M.G.L. (2000). Surpassing the lateral resolution limit by a factor of two using structured illumination microscopy. *Journal of Microscopy*, 198(Part 2), 82–87.

Gustafsson, M.G.L., Shao, L., Carlton, P.M., Wang, C.J.R., Golubovskaya, I.N., Cande, W.Z., Agard, D.A., Sedat, J.W. (2008). Three-dimensional resolution doubling in wide-field fluorescence microscopy by structured illumination. *Biophysical Journal*, 94(12), 4957–4970.

Hubert, A., Harms, F., Juvénal, R., Treimany, P., Levecq, X., Loriette, V., Farkouh, G., Rouyer, F., Fragola, A. (2019). Adaptive optics light-sheet microscopy based on direct wavefront sensing without any guide star. *Optics Letters*, 44(10), 2514–2517.

Huisken, J. and Stainier, D.Y.R. (2009). Selective plane illumination microscopy techniques in developmental biology. *Development*, 136(12), 1963–1975.

Jemielita, M., Taormina, M.J., DeLaurier, A., Kimmel, C.B., Parthasarathy, R. (2013). Comparing phototoxicity during the development of a zebrafish craniofacial bone using confocal and light sheet fluorescence microscopy techniques. *Journal of Biophotonics*, 6(11–12), 920–928.

Karadaglić, D. and Wilson, T. (2008). Image formation in structured illumination wide-field fluorescence microscopy. *Micron*, 39(7), 808–818.

Lim, D., Chu, K.K., Mertz, J. (2008). Wide-field fluorescence sectioning with hybrid speckle and uniform-illumination microscopy. *Optics Letters*, 33(16), 1819–1821.

Lim, D., Ford, T.N., Chu, K.K., Mertz, J. (2011). Optically sectioned in vivo imaging with speckle illumination HiLo microscopy. *Journal of Biomedical Optics*, 16(1), 016014.

Neil, M.A.A., Wilson, T., Juskaitis, R. (1997). Methods of obtaining optical sectioning by using structured light in a conventional microscope. *Optics Letters*, 22(24), 1905–1907.

Ruckstuhl, T., Enderlein, J., Jung, S., Seeger, S. (2000). Forbidden light detection from single molecules. *Analytical Chemistry*, 72(9), 2117–2123.

Schermelleh, L., Ferrand, A., Huser, T., Eggeling, C., Sauer, M., Biehlmaier, O., Drummen, G.P.C. (2019). Super-resolution microscopy demystified. *Nature Cell Biology*, 21(1), 72–84.

Tanaami, T., Otsuki, S., Tomosada, N., Kosugi, Y., Shimizu, M., Ishida, H. (2002). High-speed 1-frame/ms scanning confocal microscope with a microlens and nipkow disks. *Applied Optics*, 41(22), 4704–4708.

2

Contrast-based Label-Free Imaging and Phase Measurement

Pierre BON
XLIM UMR 7252, CNRS, University of Limoges, France

2.1. Introduction, the biological object as an index object

Biological samples mainly consist of water (60 to > 99%), proteins, lipids and sugars (including carbohydrates, nucleic acids, etc.). Some cells that make up biological tissues absorb light, such as, for example, retinal pigment cells or plant cells capable of photosynthesis, but the vast majority of cells exhibit very low absorption at wavelengths within the visible and near-infrared spectra. This ensures that they do not incur damage by natural light (outside ultraviolet radiation).

Nonetheless, this evolutionary advantage raises many challenges for optical imaging. The contrast observed in light transmission microscopy with a low fluence excitation (one-photon excitation) is therefore very low, especially on fine samples such as unmodified adherent cells, as shown in Figure 2.1(a).

However, the fact that light is not absorbed does not mean that the properties of the sample are not modified. The spatial distribution of the sample dry mass (namely anything that is not water) creates an equivalent distribution of the refractive index of the sample (Feynman et al. 1963). This variation in the refractive index between the cell and its medium is about 10^{-1} at most. From an optical point of view, a cell is therefore semi-transparent and can be seen as a "refractive index object". Consequently, the light is phase-shifted when it travels through the sample by a quantity that increases when the quantity of material traveled increases locally.

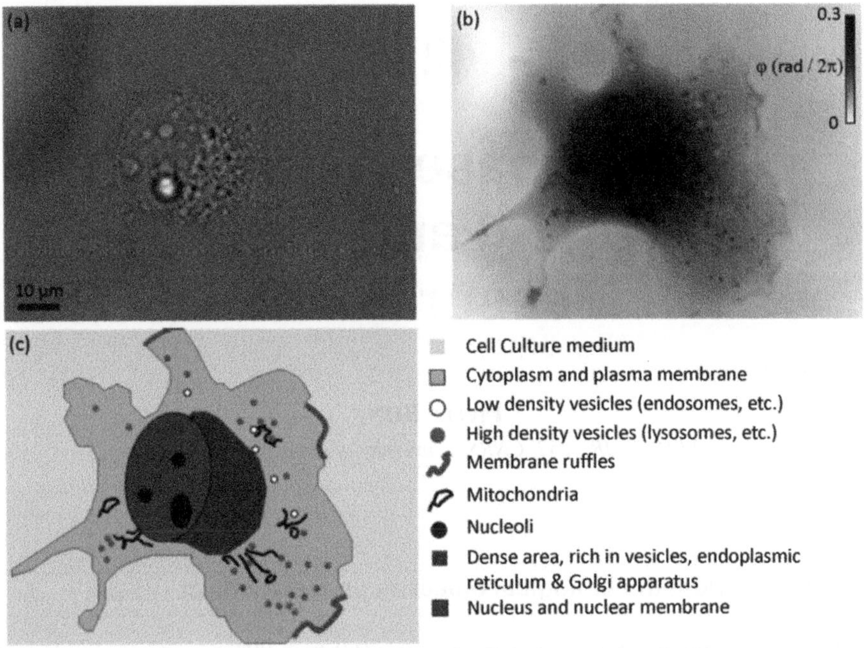

Figure 2.1. *Picture of an adherent cell (COS-7) observed under an optical microscope (100 x NA = 1.3) in the visible range (λ = 600 nm). (a) Picture of light intensity; (b) picture of the phase shift introduced by the cell; and (c) illustration of the different subcellular elements. For a color version of this figure, see www.iste.co.uk/sibarita/photonic.zip*

Therefore, if the same adherent cell is observed but this time focusing on the phase shift introduced in light (see Figure 2.1(b)), it can be seen (i) that the contrast is clearly increased compared to the picture of intensity alone (see Figure 2.1(a)) and that (ii) the phase shift is strongest at the center of the cell, where the thickness of the sample is greatest and where the material is concentrated, particularly in the nucleoli (see Figure 2.1(c)).

The phase shift will therefore carry important information about the biological sample observed. This can provide a source of contrast, on the one hand, as well as a possibility for quantifying biophysical parameters, on the other hand, due to the strong link between the phase shift and the quantity of matter (Barer 1952; Aknoun et al. 2015). As such, accounting for the phase component of light is an essential avenue for imaging in light microscopy.

In this chapter, we will describe in detail the two most widely used phase-based contrast modalities, namely Zernike phase contrast and differential interference contrast. A comparison of the different imaging modalities can be found in Figure 2.2. We will then look at the other common approaches and conclude the discussion with a description of some modalities of quantitative measurements of the phase of light.

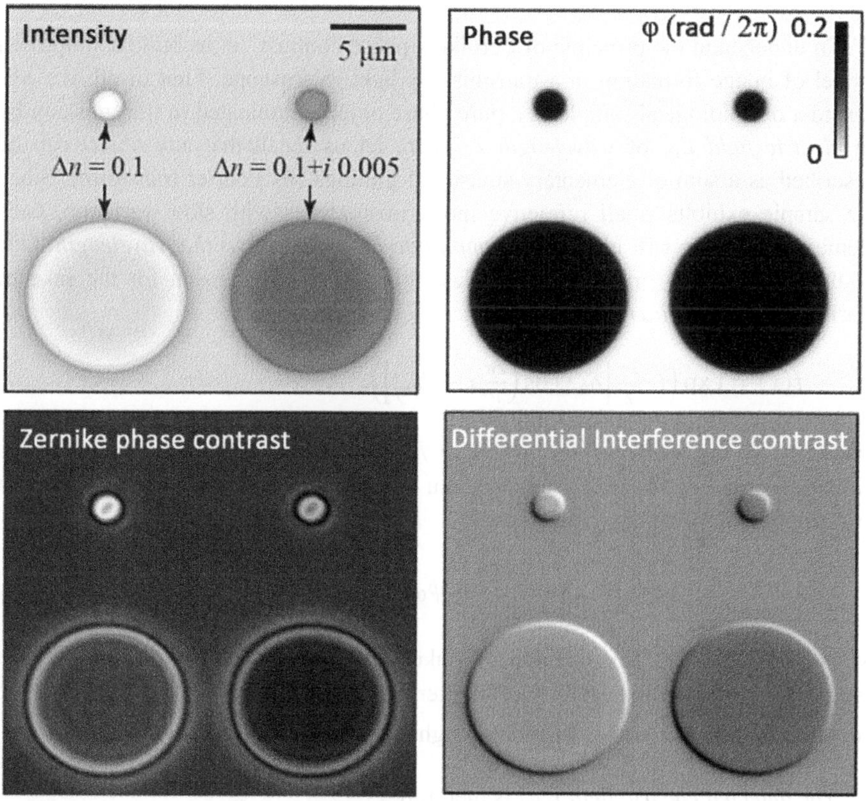

Figure 2.2. *Simulation comparison (Bon et al. 2012b) of intensity, phase, Zernike phase contrast and differential interference contrast imaging. The objects are pillars of real refractive index (phase object on the left in the pictures) and complex refractive index also reflecting absorption (on the right in the pictures) of two different sizes (a diameter of 2 μm in the top pictures and a diameter of 10 μm in the bottom pictures)*

2.2. Zernike phase contrast

As we saw in the introduction, the phase shift of light in light microscopy carries crucial information about the sample. However, to be useful, this information has to be converted into intensity modulation, i.e. into a quantity that can be measured with a detector (eye, photographic film, digital sensor, etc.).

In 1933, Frederik Zernike proposed what he called the "phase contrast" approach, which enabled the direct observation by the eye of unlabeled living cells with high contrast. This discovery quickly became essential in light microscopy, and Zernike was awarded the Nobel Prize in Physics in 1953.

To understand the principle of Zernike's phase contrast, let us build a simplified model of image formation in a transmission light microscope. First of all, we will consider our biological sample as a pure phase object illuminated in transmission by an *electric field* E_{ill} of *wavelength* λ. Then, let us recall that any object can be described as a sum of elementary sinusoidal gratings (its Fourier transform). Since the sample exhibits small refractive index mismatches with slow variation, each elementary grating will have *a low amplitude* δ_0 *compared to the wavelength*. One of these elementary gratings can be described (in one dimension for the sake of clarity) by its *(complex) transmittance* t:

$$t(x) = \exp\left(i \cdot \frac{2\pi}{\lambda}\left[\delta_0 \cdot \cos\left(\frac{2\pi}{P}x + \phi_0\right)\right]\right),$$

where ϕ_0 *is the phase at the origin of the grating*, and P *is its period*. For a biological sample, $\delta_0 \ll \lambda$ and we can simply consider a first-order Taylor expansion to describe the transmittance:

$$t(x) \approx 1 + i \cdot \frac{2\pi}{\lambda}\left[\delta_0 \cdot \cos\left(\frac{2\pi}{P}x + \phi_0\right)\right] = 1 + i \cdot \Psi(x),$$

where $|\Psi(x)| \ll 1$ is the component linked to the light beam scattered by the sample. It should be observed that the latter is phase-shifted by $\frac{\pi}{2}$ with respect to the non-scattered portion of the transmitted light.

The *detected electric field* E_{det} is therefore:

$$E_{det} = E_{ill} \cdot t \approx E_{ill} + i \cdot E_{ill} \cdot \Psi(x) = E_{ill} + E_{obj}.$$

With our model, image formation can be described as a two-wave, E_{ill} and E_{obj}, interferometer with E_{ill} being the reference beam corresponding to the light that has not been scattered by the sample, and E_{obj} being the beam corresponding to the light scattered by the object of interest (the "signal"). The image that is formed I can therefore be described as:

$$I = E_{det} \cdot E_{det}^{*} = E_{ill}^{2} \cdot \left(1 + \Psi^{2}(x)\right).$$

Since $|\Psi(x)| \ll 1$, the image will be only very slightly modulated by the object.

The principle of Zernike phase contrast consists of optimizing the contrast of this two-wave interferometer (E_{ill} and E_{obj}) in order to intensify, in particular, the modulation $\Psi(x)$ introduced by the object.

This is achieved by way of two complementary methods: (i) attenuating the reference wave after traveling through the sample to give it an energy of the same order of magnitude as that of the diffracted wave; (ii) introducing an additional phase shift on the reference wave (still after crossing the sample) to amplify the interference phenomenon. We now describe this in more detail.

Let α be *the transmission of the non-scattered field component by the object* and *ϑ be the additional phase shift applied to the scattered field with respect to the non-scattered field.* We can rewrite the detected field as follows:

$$E_{det,Zernike} = \alpha E_{ill} + i \cdot e^{i\vartheta} \cdot E_{ill} \cdot \Psi(x).$$

Therefore, the image formed becomes:

$$I_{Zernike} = E_{det,Zernike} \cdot E_{det,Zernike}^{*}$$

$$= E_{ill}^{2} \cdot \left(\alpha^{2} + \Psi(x)^{2} + 2\alpha \cdot \Psi(x) \cdot \cos\left(\frac{\pi}{2} + \vartheta\right)\right),$$

And if we choose $\vartheta = -\frac{\pi}{2}$ to compensate for the diffraction phase shift (assuming that we have a low-amplitude phase object):

$$I_{Zernike} = E_{ill}^{2} \cdot \left(\alpha^{2} + \Psi(x)^{2} + 2\alpha \cdot \Psi(x)\right) \approx E_{ill}^{2} \cdot \left(\alpha^{2} + 2\alpha \cdot \Psi(x)\right).$$

Knowing that the phase shift introduced by a living animal cell is approximately of $1/10$ rad (e.g. Figure 2.1(b)), if we choose $\alpha \sim \max \Psi \approx 1/10$, we obtain a maximization of the modulation of the image that brings forward the phase variations introduced by the sample in the form of intensity variations.

2.2.1. *Practical application and limitations*

As we can see in Figure 2.2, the contrast of phase objects is significantly increased by phase contrast. Nevertheless, a major issue with Zernike phase contrast is the presence of halos in the images. The strong assumption that non-scattered light only can be attenuated is generally not possible, and the alternative consists of attenuating and modulating certain parts of the angular spectrum of the *collected* light, mainly containing this non-scattered light (namely the illumination residual) as well as part of the light scattered by the object. This involves the appearance of artifacts in the form of halos around objects. Let us study what can be done to reduce these halos.

There remains one degree of freedom that has not been discussed yet: the angular spectrum of illumination. In the previous discussion, no assumption was made about the angle of the plane wave of illumination. Therefore, we can freely choose to illuminate the object from any angle. This will allow collection within the numerical aperture of the lens of other components scattered by the sample and thus modify the resolution and artifacts in the image.

Moreover, if we introduce not a plane wave but an incoherent superposition of plane waves, we will obtain a superposition of the intensity images formed by each of the plane waves forming the superposition, each carrying different halos. It is thus possible to modulate or even eliminate these artifacts in a controlled manner.

If the superposition of plane waves with pseudo-random angles gives the best results (Maurer et al. 2008), its implementation in a robust way remains difficult to achieve. The most widely used and commercially available solution today is an annular arrangement of the illumination angles, which is much simpler to align and produce. Figure 2.3 shows an example of a conventional implementation based on the phase shift and attenuation values calculated above.

However, this type of ring-based illumination still yields halos (as described above) due to the filtering of certain object spatial frequencies: some sizes of objects are more or less properly contrasted depending on the numerical aperture of the illumination/detection ring. Therefore, there are several apertures available that can be commercially found usually referred to as "Type I", "Type II" phase contrast… up to IV, the latter requiring an immersion condenser to get the necessary illumination numerical aperture. It should be noted that the higher the type, the better the high spatial frequencies (and therefore the small details) will be contrasted in the final image at the expense of the low frequencies.

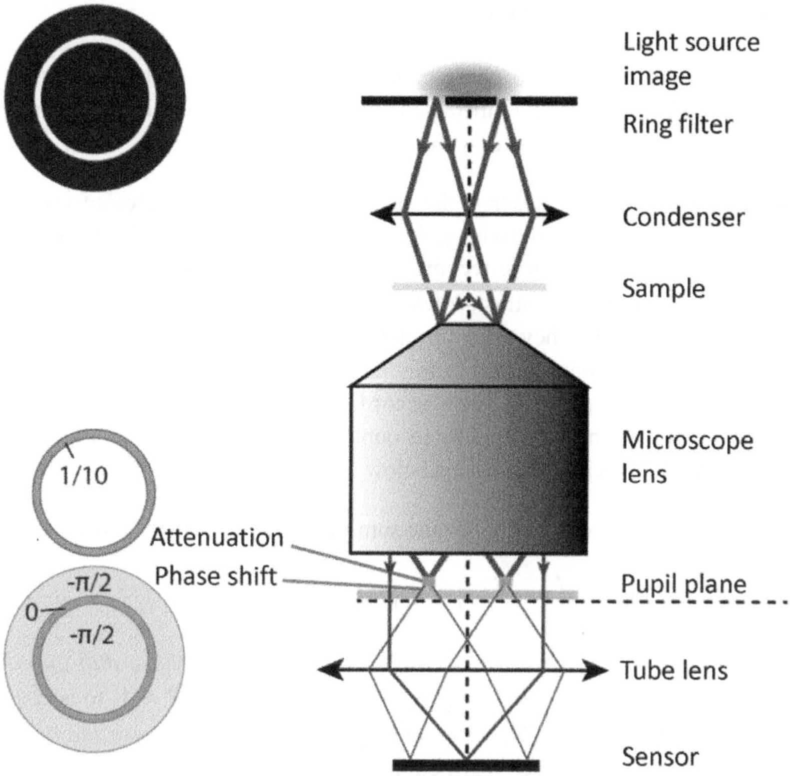

Figure 2.3. *Illustration of a Zernike phase contrast setup. For a color version of this figure, see www.iste.co.uk/sibarita/photonic.zip*

Finally, phase contrast has two other main limitations. First, it is impossible to make a distinction between a modulation of light intensity due to a phase shift from a modulation of intensity due to the absorption of the sample, which restricts the use of the Zernike phase contrast to phase objects only. Secondly, and this is what will ultimately constitute its major drawback, the presence of a physical filter *in the pupil* of the collection lens complexifies the use of phase contrast in conjunction with other imaging methods for multimodal approaches.

Indeed, the use in conjunction with fluorescence microscopy, in particular, gives poor results and some microscope manufacturers provide *external* and removable phase contrast modules for the objective, containing the phase and attenuation rings, to be removed before fluorescence imaging.

2.3. Differential interference contrast

We have seen that Zernike phase contrast is based on the optimization of the amplitude and phase of the light diffracted and non-diffracted by the sample in order to create a balanced interferometer with maximized contrast.

Another possibility to preserve the balanced side – and therefore optimal in terms of contrast – consists of dividing the beam into two coherent beams $E_{out,1}$ and $E_{out,2}$ at the sample level before recombining them before detection. In the context of differential interference contrast, *a lateral phase shift* δ between the two coherent beams is applied so that they pass through two different places, but close to the sample: this induces – on phase objects – an intensity modulation to appear in the areas where there is a phase gradient according to the direction of this shift. The value of this modulation, and therefore its contrast, will depend on the importance of this phase gradient as will be discussed below (see also Figure 2.2).

Consider the detected field E_{det} as the sum of the two coherent fields:

$$E_{det}(x) = \frac{1}{2}\left[E_{out,1}\left(x - {}^{\delta}/_{2}\right) + E_{out,2}\left(x + {}^{\delta}/_{2}\right)\right].$$

If we consider that $E_{out,1} = |E_{out,1}|e^{i\varphi}$, with φ *being the phase shift introduced by the object*, and that $E_{out,2} = E_{out,1} \cdot \beta$, with $\beta = (1-\alpha)e^{i\phi} \approx 1$, to account for a *phase shift* ϕ and a slight *difference in amplitude* $(1-\alpha)$ *between the two fields* traveling through the object, the detected intensity can be written as:

$$\begin{aligned} I_{det}(x) &= E_{det}(x) \cdot E_{det}{}^{*}(x) \\ &= \frac{1}{4}\Big[I_{out,1}\left(x - {}^{\delta}/_{2}\right) + (1-\alpha)^2 \cdot I_{out,1}\left(x + {}^{\delta}/_{2}\right) + 2 \cdot (1-\alpha) \\ &\qquad \cdot \sqrt{I_{out,1}\left(x - {}^{\delta}/_{2}\right) I_{out,1}\left(x + {}^{\delta}/_{2}\right)} \\ &\qquad \cdot \cos\left(\varphi\left(x - {}^{\delta}/_{2}\right) - \varphi\left(x + {}^{\delta}/_{2}\right) - \phi\right)\Big]. \end{aligned}$$

Assuming that the lateral shift δ is small compared to the spatial variation of the phase φ, a first-order Taylor expansion of the phase shift $\varphi\left(x \pm {}^{\delta}/_{2}\right) \approx \varphi(x) \pm {}^{\delta}/_{2} \frac{\partial\varphi(x)}{\partial x}$ is a proper approximation. We can thus rewrite the intensity as:

$$I_{det}(x) = \frac{1}{4}\Big[I_{out,1}\left(x - {}^{\delta}/_{2}\right) + (1-\alpha)^2 \cdot I_{out,1}\left(x + {}^{\delta}/_{2}\right) + 2 \cdot (1-\alpha) \cdot \sqrt{I_{out,1}\left(x - {}^{\delta}/_{2}\right) I_{out,1}\left(x + {}^{\delta}/_{2}\right)} \cdot \cos\left(\delta \frac{\partial\varphi(x)}{\partial x} - \phi\right)\Big].$$

Assuming (i) that the observed object is a phase object that does not significantly modify the intensity, it can be written at the 0th order that $I_{out,1}\left(x - {}^{\delta}/_{2}\right) \approx I_{out,1}\left(x + {}^{\delta}/_{2}\right) \approx I_{out,1}(x)$, and (ii) that the difference in amplitude α between the two fields is negligible (i.e. $\alpha \to 0$), the detected light intensity becomes:

$$I_{det}(x) = \frac{I_{out,1}(x)}{2}\left[1 + \cos\left(\delta\frac{\partial\varphi(x)}{\partial x} - \phi\right)\right].$$

It is thus observed that the signal is directly modulated by the *phase gradient* introduced by the object in the direction of the shift. In case the goal is to maximize sensitivity to the phase gradient, the shift δ acts as a lever arm but must remain small compared to the microscope resolution to avoid a blurring effect of the images. The other method for maximizing sensitivity is to play with the relative phase shift ϕ between the two replicas (also called *bias*): it can be seen that for a value of $\phi = {}^{\pi}/_{2}\,[\pi]$, the cosine of the previous equation becomes a sine and therefore, at the first order in terms of phase gradient, the light intensity can be written as:

$$I_{det}(x) = \frac{I_{out,1}(x)}{2}\left[1 \pm \delta\frac{\partial\varphi(x)}{\partial x}\right].$$

2.3.1. *Practical application and limitations*

Differential interference contrast is therefore based on the principle of multi-lateral shearing interferometry. Its actual implementation in microscopy relies on taking advantage of the polarization of light. Both coherent replicas at the origin of the interferences are the result of the recombination of two orthogonal polarizations, shifted from each other at the sample level. Spatial shift and recombination are made possible through the use of two anisotropic prisms to form Wollaston and Nomarski prisms. These prisms require the control of the light polarization *before* but also *after* the sample. The concept dates back to Jamin in 1868 (Jamin 1868) and was adapted to microscopy by Lebedeff in 1930 (Lebedeff 1930). Finally, it was Nomarski in the 1950s (Nomarski 1955) who devised the most robust and still most widely used system to date for differential interference contrast generation. The invention of the Nomarski prism made it possible to spatially shift the position of the prism along the optical axis to physically place it outside the microscope objective, where a Wollaston prism would be physically placed in the plane of the pupil of the lens (and therefore physically in the objective). The optical diagram of Figure 2.4 shows one way to achieve a conventional differential interference microscope. The polarization states and wavefronts of the aftershocks are shown on the left side of the figure.

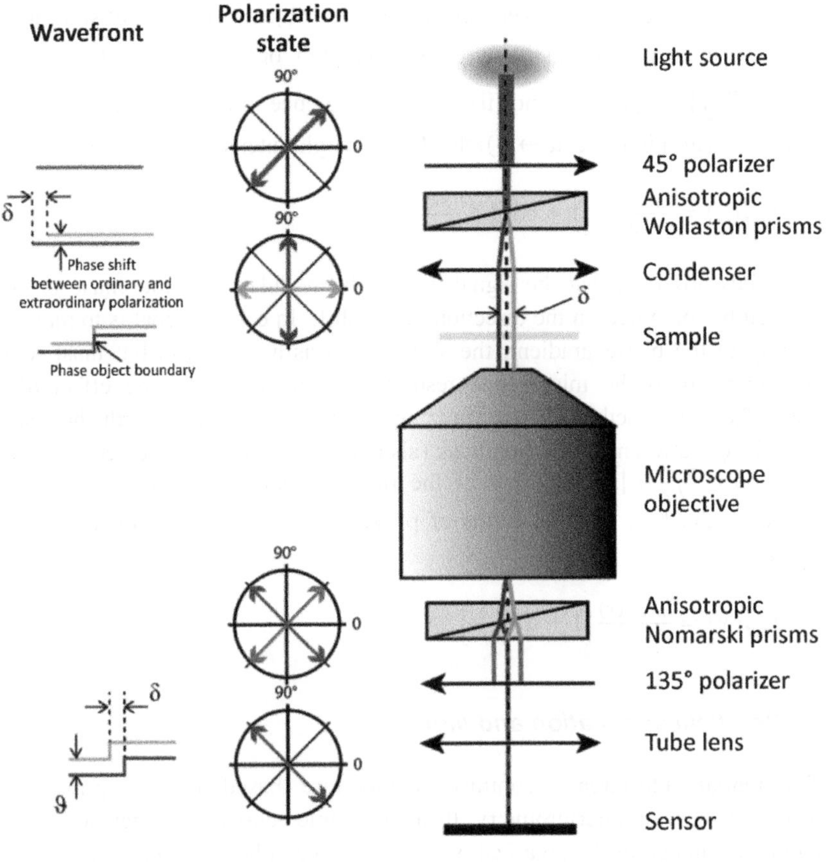

Figure 2.4. *A differential interference setup. For a color version of this figure, see www.iste.co.uk/sibarita/photonic.zip*

The differential interference contrast technique is robust when samples have low polarizing effect and low absorption. It is suitable for most sample types, and more specifically for biological samples. In particular, on thick samples, the advantage of the differential interference contrast is to provide images with very good optical sectioning because the phase gradients (source of contrast) quickly tend to zero when defocusing occurs. Interference contrast also has the advantage of not creating artifacts – as they appear with the Zernike phase contrast – allowing it to be more efficient even with relatively dense samples. In addition, it is possible to use objectives of any numerical aperture – especially those with a very high numerical aperture – the only parameter to be adjusted being the shift at the sample level with respect to the resolution of the objective being used (i.e. the choice of Wollaston prisms at excitation). This makes the technique more versatile than the Zernike contrast, and compatibility with fluorescence imaging is good if the fluorescence signal level is strong and the lateral shift is low. The presence of the polarization analyzer divides the flux of fluorescence photons (generally non-polarized) by two, and it is therefore recommended to at least remove the polarization analyzer before fluorescence imaging.

Nevertheless, this contrast technique has a number of limitations. First, the use of light polarization control during excitation and detection assumes that no other element in the light path alters the polarization states. This thus implies limiting the use of plastic sample holders and caution when using dichroic slides (especially in reflection), on the one hand, and restricting the observation to samples with very little polarization activity, on the other hand. Some manufacturers (Danz et al. 2004) provide interference contrast systems that require polarization control only after the sample at the cost of lower stability and greater alignment sensitivity.

Another limitation – just as with Zernike's phase contrast – resides in its inability to differentiate between variations in light intensity due to phase gradients and light attenuations resulting from strong absorption/scattering by the sample: this restricts its use to phase objects. Finally, its main limitation (because it concerns all samples) is caused by the lateral shift intrinsically in a single direction: the conversion of phase information into variation in light intensity is only done in this direction, and it is impossible to see the orthogonal details.

2.4. Other contrast methods with transparent objects

We have presented in detail the two most commonly used techniques in light microscopy for the observation of phase objects. There are many more – often derived from Zernike contrast or differential interference contrast – some of which are very simple to implement on a conventional microscope.

As seen for Zernike phase contrast, the presence of light not scattered by the sample represents the major contribution in terms of photons in the formation of the images. Zernike contrast seeks to weigh these energies between the scattered and non-scattered components, but a more drastic method consists of completely removing the non-scattered contribution: this is called *dark-field imaging*. The image is completely black when there is no object, and the presence of an object (with absorption and/or refractive index mismatch) will create a bright image at the level of the areas where diffraction occurs (interfaces transverse to the imaging plane). There are three main methods for addressing the dark field, (i) using a ring illumination with a numerical aperture greater than the collection numerical aperture of the microscope objective: the non-scattered light is directly rejected by the objective; (ii) placing a mask after the lens blocking non-diffracted light, a technique close to Zernike phase contrast but with zero transmission $\alpha=0$; and (iii) illuminating the sample with a given polarization and using a cross-polarization analyzer after light collection by the objective if the object is polarization-active.

The dark field allows for the detection of nano-sized objects with a very good signal-to-noise ratio. However, its use implies, as for Zernike phase contrast, that the sample be weakly scattering (i.e. thin or very transparent) in order to be able to effectively reject illumination light that does not carry useful information from the sample. In addition, the interpretation of images can prove complicated when there is a high density of elements.

The last contrast method that I would describe here is the *contrast by oblique illumination* and one of its more advanced versions, the *Hoffman modulation contrast*. The principle is to use an unbalanced illumination in terms of angular spectrum, globally oblique on one side with an average angle close to the boundaries of collection of the lens. In this case, light scattered by the sample – implicitly angularly centered around the illumination – is very anisotropic with respect to the angular spectrum collected by the imaging sample. This creates images that will look similar to differential interference contrast (the "lunar crater" aspect). The generalization of this concept – also inspired by the Foucault knife-edge technique – is called the Hoffman contrast modulation (Hoffman and Gross 1975). It uses light filtering to illuminate the sample at an angle, and filtering in the objective pupil to attenuate light that has not interacted with the sample and suppress light scattered on the same side of the pupil as the illumination angles.

Oblique illumination is extremely simple in its implementation, especially on low numerical aperture objectives, and is fully compatible with fluorescence imaging. Hoffman contrast will be more versatile in terms of numerical aperture, but the presence of the filter in the lens pupil makes its combined use – as for Zernike contrast – complicated with fluorescence.

2.5. Measuring phase quantitatively: more than just contrast

So far, we have focused on the major problem in microscopy in obtaining a contrast with so-called *phase* samples. However, apart from the simple contrast, a quantitative measurement of the phase shift introduced by the sample gives access to the quantification of quantities of biological and physical interest, such as – to name but a few – the local mass of the sample (Barer 1952), the optical refractive index of cell compounds (Bon et al. 2012a) or the temperature variation (Baffou et al. 2012).

There are many techniques for phase quantification in microscopy. Some are the generalization of Zernike phase contrasts (Wang et al. 2011) – based on a modulation of the phase ring phase shift $\boldsymbol{\vartheta}$ in the lens pupil (by means of spatial light modulators), and of the differential interference contrast (Cogswell et al. 1997) – by varying the bias $\boldsymbol{\phi}$ and the angle of lateral shift to obtain a phase gradient signal in the orthogonal direction. These quantification techniques present the same limitations in applications as the contrast-only techniques from which they are derived. Even though there are many other existing families of techniques, I will discuss two main groups below: those based on holographic interferometry and those based on lateral-shifting interferometry.

The most widely used technique for measuring the phase of light in microscopy is based on *holography*. A reference beam that has not seen the sample interferes with the beam that has traveled through the sample. When the coherence lengths are respected between both beams, they interfere and generate fringes carrying the phase and intensity information of the beam that has traveled through the sample. The most common setup is based on an off-axis Mach–Zender geometry (Cuche et al. 1999) (see Figure 2.5, left). In order to obtain a reconstruction of the phase, it requires the acquisition of only one image. This technique is relatively simple to implement and, in the end, it allows working with the scalar electromagnetic field through the joint quantitative measurement of intensity and phase. Access to the scalar electromagnetic field enables digital focusing of images (Langehanenberg et al. 2011). Nonetheless, holography faces limitations when the sample is too thick, on the one hand, with the presence of optical speckles and, on the other hand, because of phase wrapping in the $0/2\pi$ range which can lead to significant problems of interpretation.

The other family of techniques relies on making use of lateral shifting interferometry – a concept that resulted in the development of differential interference contrast. These techniques will act at the exit of the microscope by introducing an optical element (such as a diffraction grating, a diffuser, etc.) to generate coherent replicas of the incident beam and therefore be capable of interfering. We can mention quadriwave lateral shearing interferometry (Bon et al. 2009) – based on the use of a modified Hartmann-type diffraction grating (Primot and Guérineau (2000)) and described in Figure 2.5, on the right-hand side; the correlation of the speckle figure (Berto et al. 2017) – based on the use of a low-angle diffuser; or pupil partitioning – using groups of lenses in a relay plane of the microscope pupil (Parthasarathy et al. 2012).

The common point between these approaches, and related to lateral shifting interferometry, resides in the access to the phase gradient (and not the phase directly) which limits signal wrapping (as in holography) and allows imaging with greater dynamics, using any light source (as for differential interference contrast or Zernike contrast). These techniques are essentially close to interference contrast because of their sensitivity to the phase gradient. The differences are, however, as follows: (1) the measurement of several gradients in different directions, thus avoiding drops in sensitivity when using a single gradient when the phase variation follows the orthogonal direction to the gradient; (2) independent, quantitative encoding in a single intensity and phase image, thus ensuring the quantitative measurement of phase and intensity even in the presence of both phase shifting and absorbent samples.

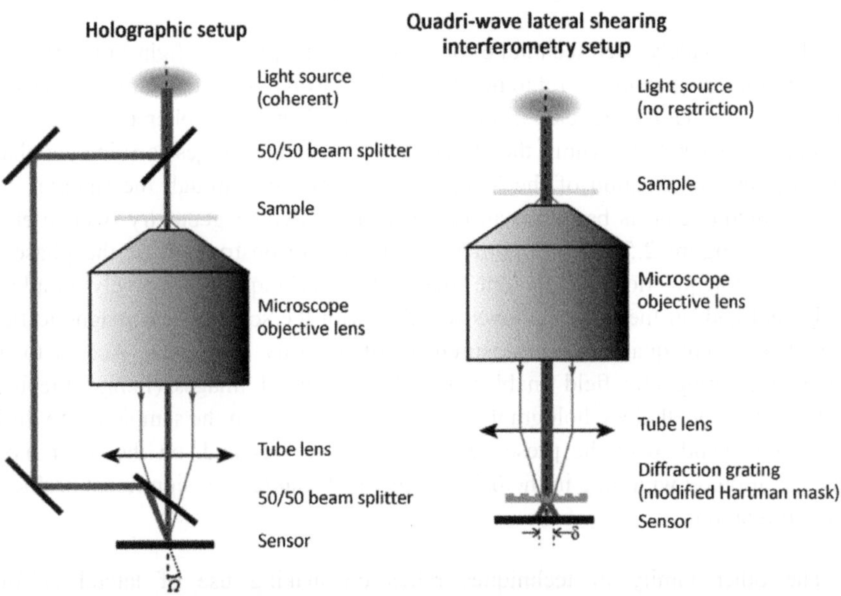

Figure 2.5. *Illustration of an off-axis holographic setup and a quadriwave lateral shearing interferometry for phase imaging setup. For a color version of this figure, see www.iste.co.uk/sibarita/photonic.zip*

However, this access to the gradient alone is problematic with flat samples if we do not see interfaces with variations: the phase measurement is thus always *relative* in the camera field of view.

2.6. Conclusions

The semi-transparent nature of biological samples has for long been the center of interest of physicists and, as we have seen in this chapter, a large number of techniques which exist today to contrast these objects and finally quantify quantities of biological interest through the use of low light powers (namely one-photon imaging).

The semi-transparent nature of biological samples, as a limiting element of contrast, proves to be a strong asset because it allows light penetration that can be centimetric in the near-infrared. Many theoretical and experimental developments are underway to challenge the boundaries of this imaging depth while maintaining the contrast necessary for the study of biological samples.

Many applications in biology and medicine are based on phase contrast-based methods in microscopy and the need to preserve biological samples, the limitation of photo-toxicity and the speed of imaging, make them cutting-edge techniques for light microscopy.

2.7. References

Aknoun, S., Savatier J., Bon, P., Galland, F., Abdeladim, L., Wattellier, B., Monneret, S. (2015). Living cell dry mass measurement using quantitative phase imaging with quadriwave lateral shearing interferometry: An accuracy and sensitivity discussion. *J. Biomed. Opt.*, 20(12), 126009.

Baffou, G. Bon, P., Savatier, J., Polleux, J., Zhu, M., Merlin, M., Rigneault, H., Monneret, S. (2012). Thermal imaging of nanostructures by quantitative optical phase analysis. *ACS Nano*, 6, 2452–2458.

Barer, R. (1952). Interference microscopy and mass determination. *Nature*, 169, 366–367.

Berto, P., Rigneault, H., Guillon, M. (2017). Wavefront sensing with a thin diffuser. *Opt. Lett.*, 42, 5117–5120.

Bon, P., Maucort, G., Wattellier, B., Monneret, S. (2009). Quadriwave lateral shearing interferometry for quantitative phase microscopy of living cells. *Opt. Express*, 17, 13080–13094.

Bon, P., Savatier, J., Merlin, M., Wattellier, B., Monneret, S. (2012a). Optical detection and measurement of living cell morphometric features with single-shot quantitative phase microscopy. *J. Biomed. Opt.*, 17, 076004.

Bon, P., Wattellier, B., Monneret, S. (2012b). Modeling quantitative phase image formation under tilted illuminations. *Opt. Lett.*, 37, 1718–1720.

Cogswell, C.J., Smith, N.I., Larkin, K.G., Hariharan, P. (1997). Quantitative DIC microscopy using a geometric phase shifter. In *Proceedings of SPIE*, 2984, 72–81.

Cuche, E., Marquet, P., Depeursinge, C. (1999). Simultaneous amplitude-contrast and quantitative phase-contrast microscopy by numerical reconstruction of Fresnel off-axis holograms. *Appl. Opt.*, 38, 6994–7001.

Danz, R., Vogelgsang, A., Käthner, R. (2004). PlasDIC – a useful modification of the differential interference contrast according to Smith/Nomarski in transmitted light arrangement. *Photonik*, 1, 42.

Feynman, R.P., Leighton, R.B., Sands, M.L. (1963). *The Feynman Lectures on Physics*. Addison-Wesley, Boston.

Hoffman, R. and Gross, L. (1975). Modulation contrast microscope. *Appl. Opt.*, 14, 1169–1176.

Jamin, J. (1868). Sur un refracteur différentiel pour la lumière polarisée. *CR Acad. Sci. (Paris)*, 67, 814.

Langehanenberg, P., von Bally, G., Kemper, B. (2011). Autofocusing in digital holographic microscopy. *3D Research*, 2, 4.

Lebedeff, A.A. (1930). L'interféromètre à polarisation et ses applications. *Rev. d'Opt*, 9, 385–413.

Maurer, C., Jesacher, A., Bernet, S., Ritsch-Marte, M. (2008). Phase contrast microscopy with full numerical aperture illumination. *Opt. Express*, 16, 19821–19829.

Nomarski, G. (1955). Nouveau dispositif pour l'observation en contraste de phase différentiel. *J. Phys. Radium*, 16(5), S88–S88.

Parthasarathy, A.B., Chu, K.K., Ford, T.N., Mertz, J. (2012). Quantitative phase imaging using a partitioned detection aperture. *Opt. Lett.*, 37, 4062–4064.

Primot, J. and Guérineau, N. (2000). Extended Hartmann test based on the pseudoguiding property of a Hartmann mask completed by a phase chessboard. *Appl. Opt.*, 39, 5715–5720.

Wang, Z. Millet, L., Mir, M., Ding, H., Unarunotai, S., Rogers, J., Gillette, M.U., Popescu, G. (2011). Spatial light interference microscopy (SLIM). *Opt. Express*, 19(2), 1016–1026.

3

Fluorophores and Labeling Methods for Fluorescence Microscopy

Jip WULFFELÉ and Dominique BOURGEOIS
Institut de Biologie Structurale, CEA, CNRS, Université Grenoble Alpes, France

3.1. Introduction

Fluorescence imaging has become an extremely powerful tool for biological studies. This is due to the unmatched sensitivity of the fluorescence physical process and the possibility to specifically label molecular targets in living samples using appropriate fluorophores. At the same time, the weakness of fluorescence microscopy is that fluorophores need to be introduced in the sample and may modify its functioning. Thus, a fluorescence-microscopy-based project always needs to start with the careful choice of fluorescent markers and labeling strategies.

This chapter provides a general introduction to fluorophores for biological imaging. It is divided into three parts: basics of photophysics, fluorescent proteins and organic dyes.

Fluorescent nanoparticles such as quantum dots are not discussed due to space limitations. For the two classes of considered fluorophores, two main aspects are described: fluorescence (so-called "photophysical") properties and labeling strategies, i.e. how to attach a fluorophore to its target. We draw special attention to fluorophore requirements specific to super-resolution methods.

Photonic Imaging for Biology,
coordinated by Jean-Baptiste SIBARITA.

3.2. Basics of fluorophore photophysics

The photophysical behavior of a fluorophore is described by the Jablonski diagram shown in Figure 3.1 (Lakowicz 2006). In the fluorescent state, the absorption of a UV–visible photon in the singlet electronic ground state S_0 generates an almost instantaneous Frank–Condon transition to a vibrationally coupled electronically excited state (S_1 or S_2), from which the fluorophore quickly relaxes (ps timescale) by internal conversion and/or vibrational relaxation to the lowest vibrational state of S_1. From there, fluorescence and non-radiative decay pathways compete (ns timescale) to bring back the fluorophore to vibrationally excited states in S_0. Alternatively, from S_1, intersystem crossing to the triplet state T_1 can occur (μs to ms timescale). T_1 is a "gate" for downstream photochemistry leading to the formation of transient (typically radical-based) dark states (ms to s timescale), or to photobleached states whereby the chromophore is irreversibly damaged.

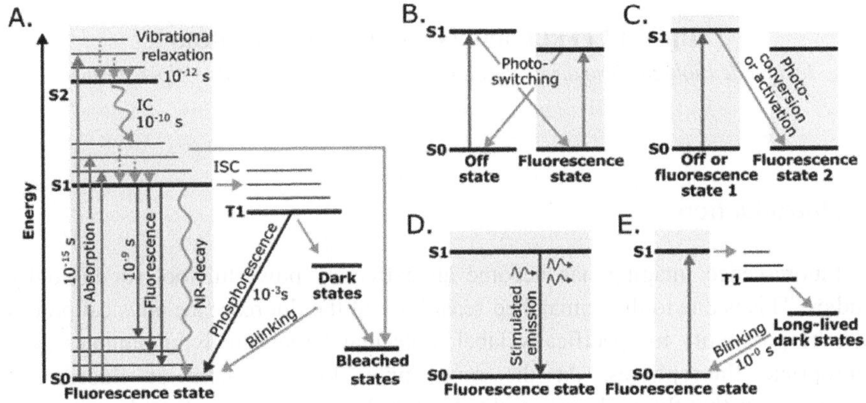

Figure 3.1. *Jablonski diagrams. (A) General case. IC: internal conversion, ISC: intersystem crossing, NR: non-radiative. (B–E) Additional transitions specific to super-resolution methods: RESOLFT (B), PALM (C), STED (D) and STORM (E). For a color version of this figure, see www.iste.co.uk/sibarita/photonic.zip*

In the fluorescent state, fluorescence brightness is defined as the product of the extinction coefficient at the excitation wavelength (efficiency of light absorption from S_0 to S_1) and the fluorescence quantum yield (probability that return to the ground state occurs through photon emission). All other pathways starting from electronically excited states can be defined by associated "phototransformation" quantum yields (probability that the pathway in question occurs). Fluorescence quantum yields often exceed 0.5, whereas phototransformation quantum yields are orders of magnitude lower (e.g. ~10^{-3} for the intersystem crossing). A key parameter is the so-called "photon budget" of a fluorophore, i.e. the total number of photons a fluorophore can emit before it photobleaches. The ns-long lifetimes of the S_1 state give rise to a strong environmental dependence of fluorescence processes. Indeed, in a matter of a few nanoseconds, many biological events (molecular diffusion, protein tumbling, solvent reorganization) can occur that might modify the energy landscape sustaining the Jablonski diagram of Figure 3.1.

At the single-molecular level, any transition to a transient dark state gives rise to an interruption of fluorescence and is referred to as "blinking". Essentially, all fluorophores blink to some extent. The nature of dark states will be discussed later in the chapter.

Super-resolution methods (Sahl et al. 2017; Liu et al. 2022) rely on specific photophysical transitions (see Figure 3.1(B–E)). Reversible photoswitching to a nonfluorescent but light-sensitive *off*-state that can be pumped back into the fluorescent state by blue-shifted illumination is at the basis of reversible saturable optical fluorescence transition (RESOLFT) microscopy.

Nonreversible photoactivation from a thermodynamically stable *off*-state to the fluorescent state, or photoconversion from one fluorescent state to another of a different color allow photoactivation localization microscopy (PALM). Enforcing return from S_1 to S_0 by stimulated emission is central to stimulated emission depletion (STED) microscopy. Stabilizing dark states so that they become very long-lived (s to min) is the concept behind stochastic optical reconstruction microscopy (STORM).

3.3. Fluorescent proteins

Fluorescent proteins (FPs) have revolutionized fluorescence microscopy (Dedecker et al. 2013). Indeed, their gene can be fused to that of a protein of interest (POI) so that, after transfection, the FP-labeled POI is expressed *in cellulo* and can be imaged. This offers ideal labeling specificity and compatibility with live-cell imaging. The huge potential of FPs for fluorescence imaging was recognized by the Nobel Prize in Chemistry in 2008 awarded to Shimomura, Chalfie and Tsien. Shimomura was the first to isolate the green fluorescent protein (GFP) from the jellyfish *Aequora victoria* in 1962 (Shimomura 2009). Chalfie later demonstrated how GFP can be used as genetic label for biological research (Chalfie et al. 1994), and Tsien extended the FP color palette and contributed to our mechanistic understanding of GFP fluorescence (Tsien 1998). Their research, the discovery of FPs in different marine organisms (hydrozoan species such as jellyfishes and anthozoan species such as corals and anemones) (Matz et al. 1999) as well as extensive mutagenesis (Shu et al. 2006) have transformed FPs into the indispensable biological tool they are nowadays.

All GFP-like FPs have a molecular weight of about 30 kDa and share a similar structure consisting of a beta-barrel with a central helix which accommodates the (p-hydroxybenzylidene-imidazolinone) chromophore that is formed by three amino acids (S65, Y66 and G67 for wtGFP) (see Figure 3.2(A)). The beta-barrel facilitates maturation of the chromophore and, after maturation, stabilizes it in position by van der Waals interactions, π-stacking interactions and an extensive hydrogen bond network, favoring fluorescence over non-radiative decay. By mutating the chromophore and/or the protein scaffold, the engineering of GFP gave rise to FPs with a variety of photophysical behaviors and covering the entire visible spectrum. The rational design of FPs with specific properties, however, remains challenging as the effect of a mutation is difficult to predict and often depends on the presence or absence of other mutations. Other types of FPs (non-GFP-like) also exist, such as FPs derived from bacteriophytochromes, reaching into the infrared range (Chernov et al. 2017). These FPs, however, require exogenous chromophores and often suffer from a low fluorescence quantum yield. Thus, the quest for new FPs continues, searching for FPs that are brighter, redder and more photostable. Below, we discuss the main properties, advantages and drawbacks of FPs that need to be considered when choosing an FP as the fluorescent label (see Figure 3.2).

Proper folding and maturation of the FP are needed for a high labeling efficiency. Misfolding of FPs may occur (as for any protein), notably under conditions that differ from those in which the (parent) FP is naturally expressed. A particular problem is the temperature difference between the cold environments in which FP-expressing organisms typically live versus the 30–37°C many bacteria and mammalian cells require. Maturation of the chromophore is a slow process which can take minutes to hours causing a delay in the appearance of fluorescence (Balleza et al. 2018).

Thus, to increase the labeling efficiency, efforts have been made to engineer FPs with optimized folding and maturation in bacterial and mammalian cells, resulting in variants such as αGFP (Crameri et al. 1996) or the well-known superfolder GFP (sfGFP) (Pédelacq et al. 2006).

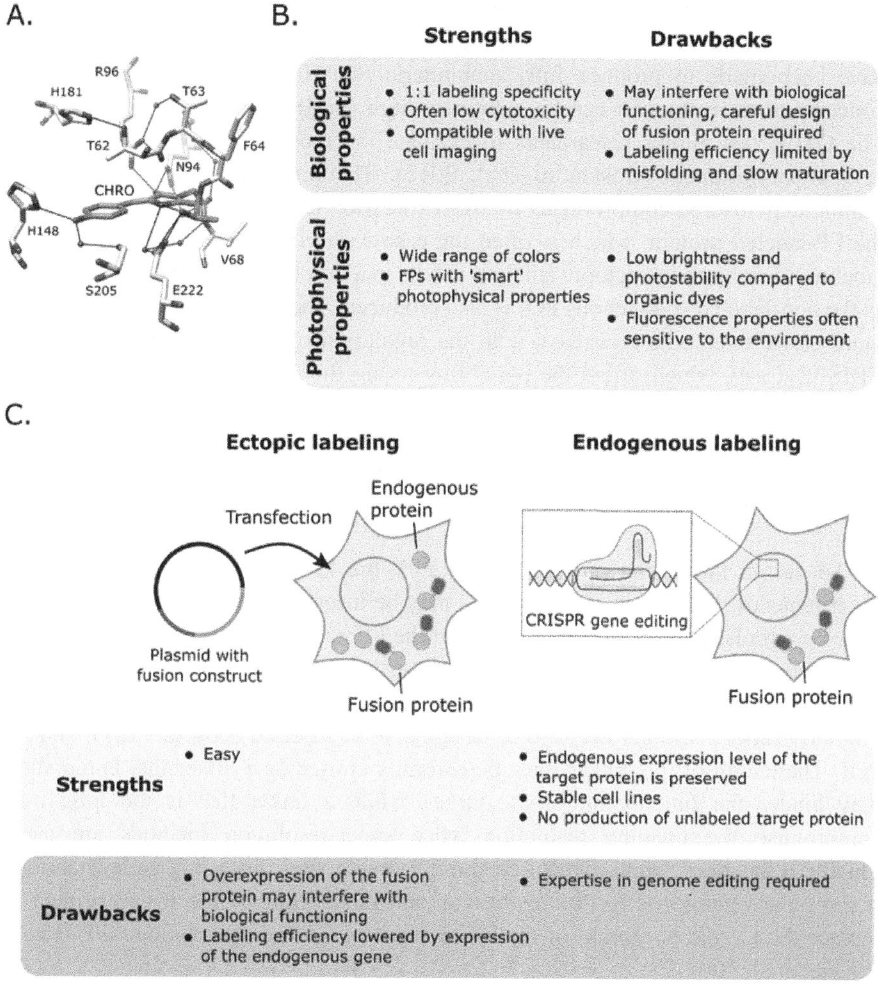

Figure 3.2. *Labeling with FPs. (A) Structure and H-bond network (black lines) of the GFP chromophore (green) with nearby residues (gray) in the β-barrel. (B) Biological and photophysical strengths and drawbacks of FPs. (C) Strengths and drawbacks of labeling strategies with FPs. For a color version of this figure, see www.iste.co.uk/sibarita/photonic.zip*

For biological studies, it is critical that the labeling does not disturb the functioning of the labeled protein. This may occur because FPs are of substantial size (30 kDa) and may block interactions of the POI with other partners or ligands. Careful design of the fusion protein is therefore needed to avoid this issue. Another problem arises from the tendency of FPs to form oligomers, especially when working with anthozoan FPs, which are obligate tetramers. Although large efforts have been made to produce fully monomeric FPs (Campbell et al. 2002), local concentrations in the cell can be extremely high (mM) and favor oligomerization. The OSER test allows researchers to evaluate the monomericity of an FP in the cellular environment (Costantini et al. 2012). The biological functioning of the sample may also be compromised by overexpression (or unregulated expression) of the FP-labeled protein, which is often the case with classical transfection (ectopic labeling). Furthermore, ectopic labeling results in a non-complete labeling efficiency as the non-labeled endogenous POI is also produced. These last two problems can be more easily overcome nowadays with the revolution of genome editing, based on CRISPR–Cas9, which offers the possibility to tag the endogenous gene, providing stable cell lines with up to 100% of the target protein labeled (case of homozygous edited cell lines) and preserving biological expression levels. Of note, to ensure optimal expression, codon optimization might be required, depending on the organism in which the FP is to be expressed.

To design the fusion protein (POI + FP), the FP can be placed at the C- or N-terminus of the POI, or in some cases may be inserted in the middle of the target gene by circular permutation. Placement of the FP needs to occur in such a way that the functioning of the target is optimally preserved. Factors to take into account, for example, are accessibility for ligands or interacting partners, and proper oligomerization. A linker needs to be designed to be inserted between the FP and the POI. The length of this linker must be carefully chosen as a linker that is too short may hinder the functioning of the target, while a linker that is too long may compromise the imaging resolution when super-resolution methods are used. Finally, it has been found that for certain constructs, typically using anthozoan FPs, it can be advantageous for the expression and functioning of the fusion protein to replace the C- and N-termini of the FP by those of GFP, a trick called GFP-ization (Shaner et al. 2004).

Brightness and photostability are key features of fluorophores. FP brightness peaks in the yellow range where the best combination of a high extinction coefficient and a high fluorescence quantum yield is found. An important factor affecting the apparent brightness of FPs is the protonation state of the chromophore: neutral (protonated at the hydroxybenzylidene oxygen atom, typically nonfluorescent) or anionic (unprotonated at the oxygen atom, typically fluorescent). As a consequence, FP fluorescence is often sensitive to pH. In fact, the brightness of FPs is dependent on various environmental factors such as ionic strength or redox potential, which is often regarded as a drawback. However, this sensitivity can also be used to engineer FP-based biosensors (Dedecker et al. 2013).

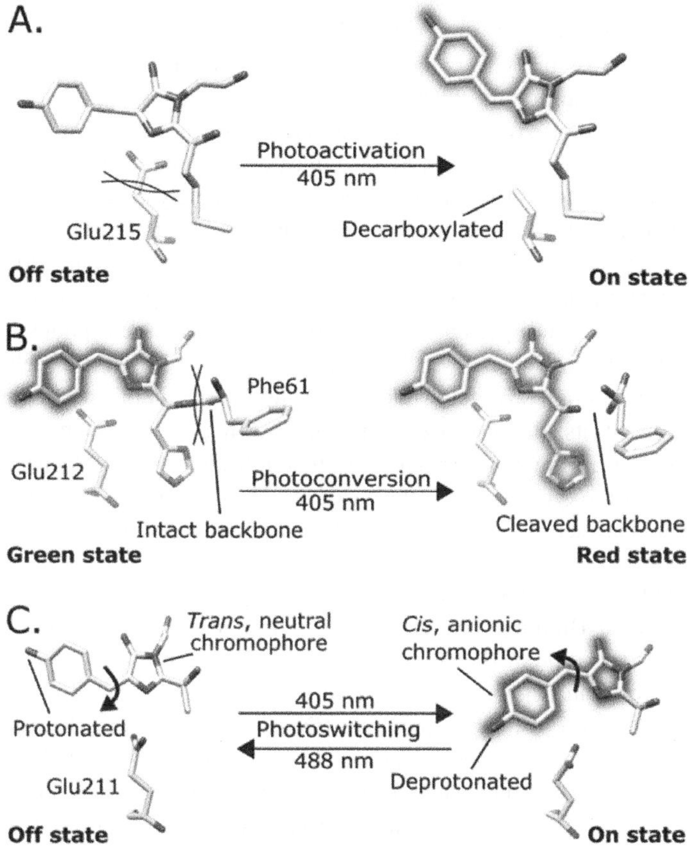

Figure 3.3. *Phototransformations in FPs. (A) Photoactivation of PAmCherry is induced by the absorption of a 405 nm light, which causes decarboxylation of Glu222 leading to a rearrangement of the hydrogen bond network stabilizing the chromophore in a deprotonated state. (B) Photoconversion of mEos4b is induced by the absorption of a 405 nm light in the green state, which causes breakage of the peptide backbone near the chromophore, extending the conjugated electron system and leading to a red shift in the absorption and emission spectra. (C) Photoswitching of rsEGFP2 is controlled by absorption of 405 nm light (dark -> fluorescent) and 488 nm light (fluorescent -> dark), which induce cis–trans isomerization and protonation–deprotonation of the chromophore. For a color version of this figure, see www.iste.co.uk/sibarita/photonic.zip*

FPs, unfortunately, are typically far less bright than organic fluorophores. This is partly explained by the large difference in the extinction coefficient between FPs and organic dyes (e.g. 55,000 $M^{-1}cm^{-1}$ for EGFP compared to 270,000 $M^{-1}cm^{-1}$ for Alexa 647). But the main reason why FPs give a lower signal is due to their relatively poor photostability as compared to organic dyes (Cranfill et al. 2016). In fact, this may seem surprising given that the FP chromophore, hidden inside the β-barrel matrix, could be expected to be better protected from the environment. Yet, the photobleaching quantum yield of FPs is typically one order of magnitude higher than that of organic dyes. How FPs bleach is incompletely understood, but it is known that it occurs by several mechanisms (Duan et al. 2013). The triplet state plays an important role in photobleaching processes for all fluorophores. In FPs, the T_1 lifetime is particularly long (ms) (Byrdin et al. 2018), giving more time for deleterious reactions and possibly at the origin of increased photobleaching. Reaction of T_1 with oxygen – which can generally still diffuse through the FP β-barrel – produces singlet oxygen, which may cause oxidation of some amino-acids and destruction of its fluorescence properties. At high laser intensities, T_1 or downstream dark states may absorb another photon, leading to photobleaching even in the absence of oxygen (Duan et al. 2013). Thus, a supra-linear relation is generally found between illumination intensity and photobleaching rate.

FPs typically can access multiple dark states, giving rise to highly complex blinking behaviors. Blinking can occur due to triplet state intersystem crossing followed by the formation of short-lived or long-lived radical states, according to the mechanisms shown in Figure 3.1. In certain FPs, blinking may also arise from *cis–trans* isomerization of the chromophore (De Zitter et al. 2019). These blinking behaviors can be strongly dependent on environmental or light illumination conditions. For example, radical state formation depends on the redox potential, and some dark states are light-sensitive, meaning that their lifetime can be reduced by suitable illumination.

Super-resolution techniques rely on the specific photophysical properties of so-called "phototransformable" FPs, whose photophysics can be controlled by illumination. Three types of phototransformable FPs are distinguished: photoactivatable, photoconvertible and photoswitchable FPs (see Figure 3.3) (Adam et al. 2014). Photoactivatable FPs (PAFPs) can be switched irreversibly from a dark state to a fluorescent state by illumination with violet light. Photoactivation typically involves a change in the protonation state of the chromophore induced by decarboxylation of Glu222 (GFP numbering, a fully conserved residue located near the chromophore), favoring the fluorescent anionic state. Photoconvertible FPs (PCFPs) can be irreversibly switched from one color to another, which offers an advantage over PAFPs as both states are observable by fluorescence microscopy. Green-to-red PCFPs were discovered in anthozoan species. The underlying mechanism of photoconversion is based on an elongation of the conjugated electron system that results from a violet light-induced backbone breakage, but the details of this mechanism remain poorly understood. PCFPs display incomplete photoconversion (50–70%) which poses challenges for PALM experiments (Durisic et al. 2014). Popular PAFPs and PCFPs are PAmCherry (Subach et al. 2009) and the mEos variants (Nienhaus and Nienhaus 2021), respectively. In contrast to PAFPs and PCFPs, the phototransformation of reversibly switchable FPs (RSFPs) is reversible as they can be switched back and forth between a dark and a fluorescent state. These FPs are used for nonlinear structured-illumination microscopy (NL-SIM) and RESOLFT microscopy. The switching of RSFPs is most often based on *cis–trans* isomerization of the chromophore, which is possible because in these proteins the chromophore is less restricted (Bourgeois and Adam 2012). Unfortunately, this also means that RSFPs have a relatively low fluorescence quantum yield, especially the fast-switching ones. An example of a popular RSFP today is rsEGFP2 (Grotjohann et al. 2012).

In summary, the 100% labeling specificity and compatibility with live-cell imaging make FPs the preferred label for many microscopy experiments. However, FPs still need to be improved, notably in terms of brightness and photostability.

3.4. Organic dyes

In parallel with fluorescent proteins, there exists a huge plethora of organic fluorophores and fluorescent nanoparticles that are used for diverse applications in fluorescence imaging. Intense research continues to make those fluorophores increasingly performant, and extensive reviews have been published recently to highlight advances (Lavis and Raines 2008; Jin et al. 2018; Li and Vaughan 2018). Here, we concentrate on the basic features about organic dyes. Figure 3.4 summarizes the main advantages and drawbacks of these fluorophores, which can be compared to those of fluorescent proteins described above.

Classifying organic fluorophores is a difficult exercise, but it is possible to distinguish five main subfamilies:

– naturally occurring fluorophores such as nicotinamide adenine dinucleotide (NADH), flavins or some amino acids such as tryptophan, phenylalanine or tyrosine;

– polycyclic aromatics (e.g. pyrene), oxygen- or nitrogen-substituted bicyclic aromatics (e.g. coumarins or quinine, a well-known and historical fluorophore found in gin and tonic), indoles (e.g. DAPI or Hoechst, known for their ability to stain DNA). Typically, these fluorophores absorb and emit in the UV-blue region of the spectrum;

– xanthene-derived fluorophores: this is the widest subfamily, starting with the well-known fluorescein and fluorescein isothiocyanate (FITC), and also all the rhodamine derivatives, including sulfo-rhodamine, carboxy rhodamine, silicon rhodamine and boron-substituted xanthenes (Bodipy dyes). They cover the whole visible range up to the near-IR;

– polymethine-based fluorophores, referred to as cyanine dyes. Depending on the number of methines, they also cover the visible range up to the near-IR;

– oxazine dyes, which are tri-aromatics bearing a nitrogen and an oxygen atom on the central ring, typically fluorescing in the red region of the spectrum.

The three subfamilies of organic fluorophores mainly used in super-resolution fluorescence microscopy are rhodamine, cyanine and oxazine dyes, whose chemical structures are shown in Figure 3.4(A). We note in passing that the above-mentioned subfamilies of organic fluorophores should not be confused with their brand names. For example, Atto 488, Alexa 555 and Janelia Farm (JF) 549 are all rhodamines, whereas Alexa 488 and Alexa 647 are rhodamine and cyanine dyes, respectively. Thus far, rhodamine-based dyes have been most amenable to chemical engineering. One can cite the design of JF dyes bearing azetidine rings for improved brightness (Grimm et al. 2015), or deuteration procedures for improved photon budget (Grimm et al. 2021). Caged rhodamine dyes have also been devised to confer photoactivation properties, essential for approaches such as single-particle tracking PALM (sptPALM) (Grimm et al. 2016; Frei et al. 2019).

An important field of development is that of fluorogenic dyes, which only become fluorescent once bound to their target, allowing a drastic reduction of nonspecific background signals (Kozma and Kele 2019). Fluorogenicity in rhodamines can be reached through the establishment of an equilibrium between a nonfluorescent spirolactone form (favored in a nonpolar environment) and a fluorescent zwitterionic form (favored in a polar environment, typically when bound to a substrate) such as in the case of carboxy-SiR650 (Lukinavičius et al. 2013). Many other fluorogenic dyes have been engineered such as lipid probes (Collot et al. 2018) or DNA-binding dyes (Rye et al. 1992). An elegant way to achieve fluorogenicity is to elicit fluorescence through a reduction in dye conformational flexibility upon binding a fluorogen activating protein (FAP) tag such as the protein FAST (Plamont et al. 2016).

Many developments have been carried out to optimize organic dyes for super-resolution microscopy (Li and Vaughan 2018), in particular for STED and

STORM. In STED, dye photostability and efficient singlet-state quenching by red-shifted light are key parameters (see Figure 3.1(D)). In STORM, dark states, often anionic radicals or thiol adducts, are stabilized by adding a buffer that typically contains an oxygen scavenging system such as glucose, glucose oxidase and catalase to remove oxygen and prevent triplet state quenching, as well as a highly concentrated reducing agent (e.g. mercaptoethylamine at 10–100 mM in the case of cyanine dyes) to promote the formation of the dark state (see Figure 3.1(E)). To avoid the necessity of using such a buffer, which is only compatible with fixed cells, oxazine dyes can be used, offering the prospect for live cells experiments (Vogelsang et al, 2009). Spontaneously blinking fluorophores have also been developed that naturally achieve an equilibrium between a closed (nonfluorescent) and an open (fluorescent) form (Uno et al. 2014; Takakura et al. 2017).

The use of exogenously supplied organic dyes requires the crossing of the biological membrane. Usually, this is achieved on fixed cells by membrane permeabilization using surfactants such as Triton-X100.

Alternatively, many recently developed dyes have the capacity to cross intact plasma membranes (Grimm et al. 2015). Some of these dyes are fluorogenic and penetrate the cell in a nonpolar nonfluorescent form that shifts to a polar fluorescent form once binding to their target in the cytoplasm, as is the case for carboxylated silicon-containing rhodamine (Lukinavičius et al. 2013; Xue et al. 2015). Alternatively, some dyes might be attached to, for example, cell penetrating peptides (Pan et al. 2014).

An important aspect of organic fluorophores is that, as they are not protected from the solvent unlike (to some extent) fluorescent proteins, their photophysical behaviors strongly depend on the physicochemical environment. This can be an advantage in the case of, for example, fluorogenic dyes, but it also means that a fluorophore that performs well in a specific biological environment may perform poorly in another, depending on e.g. pH or redox potential. A variety of anti-fading agents can also be added to the medium to improve photostability, and self-healing dyes have even been engineered, which are covalently linked to protective chemical groups (Altman et al. 2012).

Organic fluorophores are only valuable if they can be specifically attached to their biological target. Here, full genetically encoded labeling is not possible as with FPs, and a very diverse set of alternative labeling strategies have been developed over the years (see Figure 3.4(C)). For a comprehensive recent review, see Choquet et al. (2021).

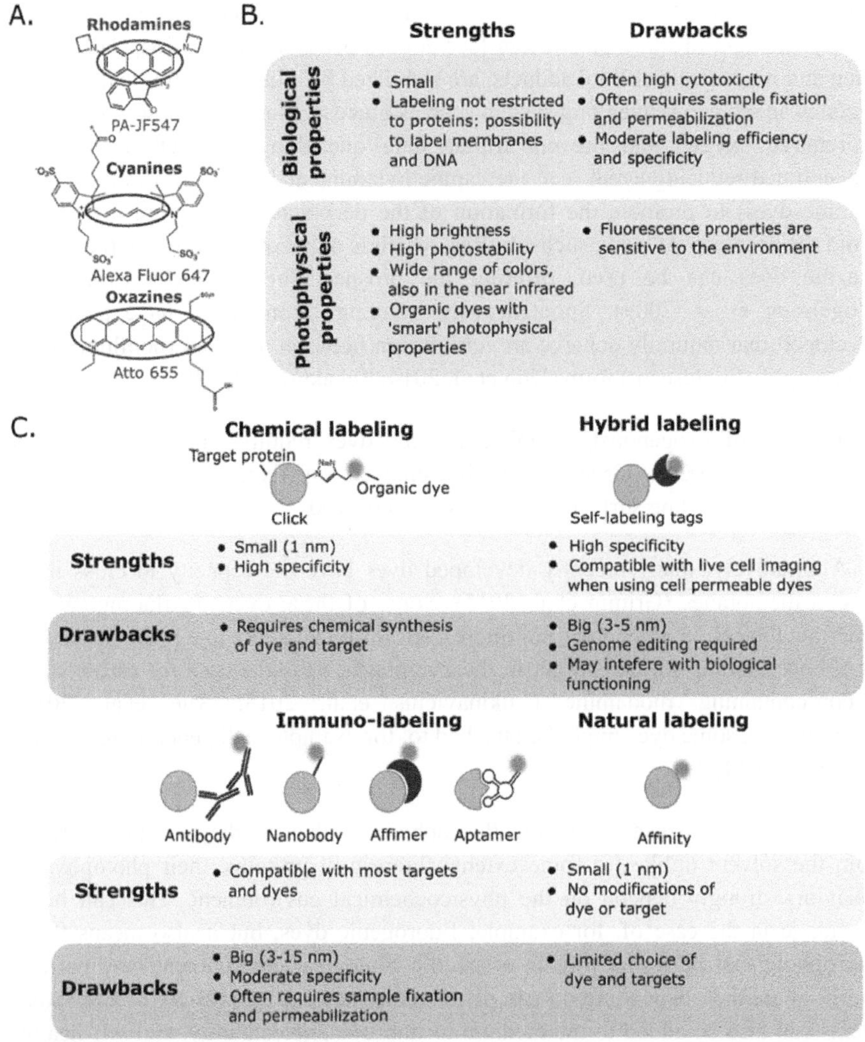

Figure 3.4. *Labeling with organic fluorophores. (A) Example structures of organic dyes. (B) Biological and photophysical strengths and drawbacks of organic dyes. (C) Strengths and drawbacks of different labeling strategies with organic dyes. For a color version of this figure, see www.iste.co.uk/sibarita/photonic.zip*

Immunolabeling is a very widespread labeling strategy. Most commonly, it is performed with antibodies, usually a primary antibody coupled to a secondary antibody to which the fluorophore is attached. The use of a pair of antibodies is justified by technical and cost considerations, as this allows using generic commercially available fluorophore-labeled secondary antibodies instead of having to design a fluorophore attachment to every target-specific primary antibody. As a consequence, the entire label is very big, reaching ~10 nm in size, which constitutes a clear limitation to the achievable resolution in nanoscopy approaches. To reduce this size, nanobodies, which are derived from single-chain antibodies produced by camelids (Hamers-Casterman et al. 1993) and have a molecular weight of ~15 kDa (about half of GFP), can be used, for example, to recognize small tags appended to the sequence of the target of interest (Virant et al. 2018).

Alternatively, aptamers are RNA-based labels that can be designed by iterative combinatorial synthesis (SELEX method) to selectively recognize a ligand with very high affinity (Ellington and Szostak 1990), while affimers are small proteins of about 12 kDa derived from cysteine proteases that can be designed by directed evolution to recognize a specific target (Woodman et al. 2005). Both aptamers and affimers are small labels appropriate for nanoscopy techniques but require specific developments for each new target.

A highly popular labeling strategy nowadays consists of generating a fusion protein made of the target of interest linked to a protein tag displaying strong affinity for an organic fluorophore. Such a strategy is called hybrid labeling and can be performed with a number of tags such as Halo-Tag (Los et al. 2008), Clip-Tag (Gautier et al. 2008), Snap-Tag (Juillerat et al. 2003) or DHFR-Tag (Wombacher et al. 2010). Binding of the fluorophore to the tag can be covalent (Juillerat et al. 2003; Gautier et al. 2008; Los et al. 2008) or noncovalent (Wombacher et al. 2010). FAPs, described above, also belong to the class of hybrid-labeling tags. In contrast to, for example, immunolabeling, but similar to labeling with fluorescent proteins, hybrid-labeled targets are ectopically (over)expressed, or their endogenous expression requires genetic engineering via, for example, CRISPR–Cas9 gene insertion.

Binding a fluorophore directly to the target in a highly compact manner is possible by using chemical labeling, typically combining the expansion of the genetic code to insert an unnatural amino acid (UAA) into the target, with click chemistry to attach a fluorophore in a biorthogonal way (Plass et al. 2011). By means of a tRNA/tRNA-synthetase pair, UAAs carrying strained alkyne or alkene groups can be genetically incorporated at positions encoded by an Amber (TAG) STOP codon. Efficient labeling is then achieved with azide-bearing (fluorogenic) dyes by means of a click reaction under physiological conditions. Alternatively, click chemistry can be elegantly used to directly label an azide-bearing clickable metabolite which is added to the sample (Trouve et al. 2021).

Finally, in some cases, direct labeling can be achieved by conjugating a fluorophore to a molecule that exhibit a natural affinity to the target: for example, phalloidin is a (poisonous) molecule classically used to irreversibly label actin in fixed cells, while Lifeact is a short peptide that can transiently label actin in live cells (Riedl et al. 2008).

In summary, the brightness, photostability and color tunability of organic dyes make them preferred labels for a wide diversity of imaging experiments, notably in chemically fixed cells. However, despite significant recent progress, many organic dyes remain incompatible with live-cell experiments and may produce nonspecific staining.

3.5. Conclusion

The success of any biological fluorescence imaging experiment heavily depends on the choice of suitable fluorophores. Nowadays, there is a wide range of FPs and organic dyes available, which all have their own strengths and drawbacks. There is not one perfect fluorophore for all experiments. Instead, what is "the best" fluorophore for an experiment depends on what is to be studied: a whole organism or a single cell; a live sample or a fixed sample; at room temperature or at cryogenic temperature.

Likewise, there is still no ideal labeling strategy and every project requires its own optimization. Factors to consider are labeling specificity, labeling efficiency, preservation of biological function, compatibility with endogenous expression, compatibility with live-cell experiments and, finally, label size for compatibility with super-resolution approaches.

New developments are ongoing to improve the photophysics and labeling of fluorophores. Brightness and photostability are of key concern, especially for super-resolution techniques which often rely on the detection of single fluorophores. Labeling with FPs has been significantly improved by the invention of CRISPR gene editing, but misfolding, slow maturation and photobleaching still limit their effective labeling efficiency.

On the other hand, organic dyes, although superior to FPs in brightness and photostability, still lack the 100% labeling specificity and full compatibility with live-cell imaging in most cases. New developments lie ahead for fluorophores with improved photophysics and novel labeling strategies with higher efficiency and precision. They will provide the tools to further unravel the mysteries of life.

3.6. References

Adam, V., Berardozzi, R., Byrdin, M., Bourgeois, D. (2014). Phototransformable fluorescent proteins: Future challenges. *Current Opinion in Chemical Biology*, 20, 92–102.

Altman, R.B., Zheng, Q., Zhou, Z., Terry, D.S., Warren, J.D., Blanchard, S.C. (2012). Enhanced photostability of cyanine fluorophores across the visible spectrum. *Nature Methods*, 9, 428–429.

Balleza, E., Kim, J.M., Cluzel, P. (2018). Systematic characterization of maturation time of fluorescent proteins in living cells. *Nature Methods*, 15, 47–51.

Bourgeois, D. and Adam, V. (2012). Reversible photoswitching in fluorescent proteins: A mechanistic view. *IUBMB Life*, 64, 482–491.

Byrdin, M., Duan, C., Bourgeois, D., Brettel, K. (2018). A long-lived triplet state is the entrance gateway to oxidative photochemistry in green fluorescent proteins. *Journal of the American Chemical Society*, 140, 2897–2905.

Campbell, R.E., Tour, O., Palmer, A.E., Steinbach, P.A., Baird, G.S., Zacharias, D.A., Tsien, R.Y. (2002). A monomeric red fluorescent protein. *Proceedings of the National Academy of Sciences of the USA*, 99, 7877–7882.

Chalfie, M., Tu, Y., Euskirchen, G., Ward, W.W., Prasher, D.C. (1994). Green fluorescent protein as a marker for gene expression. *Science (New York)*, 263, 802–805.

Chernov, K.G., Redchuk, T.A., Omelina, E.S., Verkhusha, V.V. (2017). Near-infrared fluorescent proteins, biosensors, and optogenetic tools engineered from phytochromes. *Chemical Reviews*, 117, 6423–6446.

Choquet, D., Sainlos, M., Sibarita, J.-B. (2021). Advanced imaging and labelling methods to decipher brain cell organization and function. *Nature Reviews. Neuroscience*, 22, 237–255.

Collot, M., Fam, T.K., Ashokkumar, P., Faklaris, O., Galli, T., Danglot, L., Klymchenko, A.S. (2018). Ultrabright and fluorogenic probes for multicolor imaging and tracking of lipid droplets in cells and tissues. *Journal of the American Chemical Society*, 140, 5401–5411.

Costantini, L.M., Fossati, M., Francolini, M., Snapp, E.L. (2012). Assessing the tendency of fluorescent proteins to oligomerize under physiologic conditions. *Traffic*, 13, 643–649.

Crameri, A., Whitehorn, E.A., Tate, E., Stemmer, W.P.C. (1996). Improved green fluorescent protein by molecular evolution using DNA shuffling. *Nature Biotechnology*, 14, 315–319.

Cranfill, P.J., Sell, B.R., Baird, M.A., Allen, J.R., Lavagnino, Z., de Gruiter, H.M., Kremers, G.-J., Davidson, M.W., Ustione, A., Piston, D.W. (2016). Quantitative assessment of fluorescent proteins. *Nature Methods*, 13, 557–562.

De Zitter, E., Thédié, D., Mönkemöller, V., Hugelier, S., Beaudouin, J., Adam, V., Byrdin, M., et al. (2019). Mechanistic investigation of mEos4b reveals a strategy to reduce track interruptions in sptPALM. *Nature Methods*, 16, 707–710.

Dedecker, P., De Schryver, F.C., Hofkens, J. (2013). Fluorescent proteins: Shine on, you crazy diamond. *Journal of the American Chemical Society*, 135, 2387–2402.

Duan, C., Adam, V., Byrdin, M., Ridard, J., Kieffer-Jaquinod, S., Morlot, C., Arcizet, D., Demachy, I., Bourgeois, D. (2013). Structural evidence for a two-regime photobleaching mechanism in a reversibly switchable fluorescent protein. *Journal of the American Chemical Society*, 135, 15841–15850.

Durisic, N., Laparra-Cuervo, L., Sandoval-Álvarez, Á., Borbely, J.S., Lakadamyali, M. (2014). Single-molecule evaluation of fluorescent protein photoactivation efficiency using an in vivo nanotemplate. *Nature Methods*, 11, 156–162.

Ellington, A.D. and Szostak, J.W. (1990). In vitro selection of RNA molecules that bind specific ligands. *Nature*, 346, 818–822.

Frei, M.S., Hoess, P., Lampe, M., Nijmeijer, B., Kueblbeck, M., Ellenberg, J., Wadepohl, H., Ries, J., Pitsch, S., Reymond, L., Johnsson, K. (2019). Photoactivation of silicon rhodamines via a light-induced protonation. *Nature Communications*, 10, 4580.

Gautier, A., Juillerat, A., Heinis, C., Corrêa, I.R., Kindermann, M., Beaufils, F., Johnsson, K. (2008). An engineered protein tag for multiprotein labeling in living cells. *Chemistry & Biology*, 15, 128–136.

Grimm, J.B., English, B.P., Chen, J., Slaughter, J.P., Zhang, Z., Revyakin, A., Patel, R. et al. (2015). A general method to improve fluorophores for live-cell and single-molecule microscopy. *Nature Methods*, 12, 244–250.

Grimm, J.B., English, B.P., Choi, H., Muthusamy, A.K., Mehl, B.P., Dong, P., Brown, T.A. et al. (2016). Bright photoactivatable fluorophores for single-molecule imaging. *Nature Methods*, 13, 985–988.

Grimm, J.B., Xie, L., Casler, J.C., Patel, R., Tkachuk, A.N., Falco, N., Choi, H., Lippincott-Schwartz, J., Brown, T.A., Glick, B.S. et al. (2021). A general method to improve fluorophores using deuterated auxochromes. *JACS Au*, 1, 690–696.

Grotjohann, T., Testa, I., Reuss, M., Brakemann, T., Eggeling, C., Hell, S.W., Jakobs, S. (2012). rsEGFP2 enables fast RESOLFT nanoscopy of living cells. *eLife*, 1, e00248.

Hamers-Casterman, C., Atarhouch, T., Muyldermans, S., Robinson, G., Hammers, C., Songa, E.B., Bendahman, N., Hammers, R. (1993). Naturally occurring antibodies devoid of light chains. *Nature*, 363, 446–448.

Jin, D., Xi, P., Wang, B., Zhang, L., Enderlein, J., van Oijen, A.M. (2018). Nanoparticles for super-resolution microscopy and single-molecule tracking. *Nature Methods*, 15, 415–423.

Juillerat, A., Gronemeyer, T., Keppler, A., Gendreizig, S., Pick, H., Vogel, H., Johnsson, K. (2003). Directed evolution of O6-alkylguanine-DNA alkyltransferase for efficient labeling of fusion proteins with small molecules in vivo. *Chemistry & Biology*, 10, 313–317.

Kozma, E. and Kele, P. (2019). Fluorogenic probes for super-resolution microscopy. *Organic & Biomolecular Chemistry*, 17, 215–233.

Lakowicz, J. (2006). *Principles of Fluorescence Spectroscopy*. Springer, Boston, MA.

Lavis, L.D. and Raines, R.T. (2008). Bright ideas for chemical biology. *ACS Chemical Biology*, 3, 142–155.

Li, H. and Vaughan, J.C. (2018). Switchable fluorophores for single-molecule localization microscopy. *Chemical Reviews*, 118, 9412–9454.

Liu, S., Hoess, P., Ries, J. (2022). Super-resolution microscopy for structural cell biology. *Annual Review of Biophysics*, 51, 301–326.

Los, G.V., Encell, L.P., McDougall, M.G., Hartzell, D.D., Karassina, N., Zimprich, C., Wood, M.G. et al. (2008). HaloTag: A novel protein labeling technology for cell imaging and protein analysis. *ACS Chemical Biology*, 3, 373–382.

Lukinavičius, G., Umezawa, K., Olivier, N., Honigmann, A., Yang, G., Plass, T., Mueller, V. et al. (2013). A near-infrared fluorophore for live-cell super-resolution microscopy of cellular proteins. *Nature Chemistry*, 5, 132–139.

Matz, M.V., Fradkov, A.F., Labas, Y.A., Savitsky, A.P., Zaraisky, A.G., Markelov, M.L., Lukyanov, S.A. (1999). Fluorescent proteins from nonbioluminescent Anthozoa species. *Nature Biotechnol*, 17, 969–973.

Nienhaus, K. and Nienhaus, G.U. (2021). Fluorescent proteins of the EosFP clade: Intriguing marker tools with multiple photoactivation modes for advanced microscopy. *RSC Chemical Biology*, 2, 796–814.

Pan, D., Hu, Z., Qiu, F., Huang, Z.-L., Ma, Y., Wang, Y., Qin, L. et al. (2014). A general strategy for developing cell-permeable photo-modulatable organic fluorescent probes for live-cell super-resolution imaging. *Nature Communications*, 5, 5573.

Pédelacq, J.-D., Cabantous, S., Tran, T., Terwilliger, T.C., Waldo, G.S. (2006). Engineering and characterization of a superfolder green fluorescent protein. *Nature Biotechnology*, 24, 79–88.

Plamont, M.-A., Billon-Denis, E., Maurin, S., Gauron, C., Pimenta, F.M., Specht, C.G., Shi, J., Quérard, J., Pan, B., Rossignol, J. et al. (2016). Small fluorescence-activating and absorption-shifting tag for tunable protein imaging in vivo. *Proceedings of the National Academy of Sciences of the USA*, 113, 497–502.

Plass, T., Milles, S., Koehler, C., Schultz, C., Lemke, E.A. (2011). Genetically encoded copper-free click chemistry. *Angewandte Chemie International Edition*, 50, 3878–3881.

Riedl, J., Crevenna, A.H., Kessenbrock, K., Yu, J.H., Neukirchen, D., Bista, M., Bradke, F., Jenne, D., Holak, T.A., Werb, Z. et al. (2008). Lifeact: A versatile marker to visualize F-actin. *Nature Methods*, 5, 605–607.

Rye, H.S., Yue, S., Wemmer, D.E., Quesada, M.A., Haugland, R.P., Mathies, R.A., Glazer, A.N. (1992). Stable fluorescent complexes of double-stranded DNA with bis-intercalating asymmetric cyanine dyes: Properties and applications. *Nucleic Acids Research*, 20, 2803–2812.

Sahl, S.J., Hell, S.W., Jakobs, S. (2017). Fluorescence nanoscopy in cell biology. *Nature Reviews. Molecular Cell Biology*, 18, 685–701.

Shaner, N.C., Campbell, R.E., Steinbach, P.A., Giepmans, B.N.G., Palmer, A.E., Tsien, R.Y. (2004). Improved monomeric red, orange and yellow fluorescent proteins derived from *Discosoma* sp. red fluorescent protein. *Nature Biotechnology*, 22, 1567–1572.

Shimomura, O. (2009). Discovery of green fluorescent protein (GFP) (Nobel lecture). *Angewandte Chemie*, 48, 5590–5602.

Shu, X., Shaner, N.C., Yarbrough, C.A., Tsien, R.Y., Remington, S.J. (2006). Novel chromophores and buried charges control color in mFruits. *Biochemistry*, 45, 9639–9647.

Subach, F.V., Patterson, G.H., Manley, S., Gillette, J.M., Lippincott-Schwartz, J., Verkhusha, V.V. (2009). Photoactivatable mCherry for high-resolution two-color fluorescence microscopy. *Nature Methods*, 6, 153–159.

Takakura, H., Zhang, Y., Erdmann, R.S., Thompson, A.D., Lin, Y., McNellis, B., Rivera-Molina, F. et al. (2017). Long time-lapse nanoscopy with spontaneously blinking membrane probes. *Nature Biotechnology*, 35, 773–780.

Trouve, J., Zapun, A., Arthaud, C., Durmort, C., Di Guilmi, A.M., Söderström, B., Pelletier, A. et al. (2021). Nanoscale dynamics of peptidoglycan assembly during the cell cycle of Streptococcus pneumoniae. *Current Biology: CB*, 31, 2844–2856.e6.

Tsien, R.Y. (1998). The green fluorescent protein. *Annual Review of Biochemistry*, 67, 509–544.

Uno, S., Kamiya, M., Yoshihara, T., Sugawara, K., Okabe, K., Tarhan, M.C., Fujita, H. et al. (2014). A spontaneously blinking fluorophore based on intramolecular spirocyclization for live-cell super-resolution imaging. *Nature Chemistry*, 6, 681–689.

Virant, D., Traenkle, B., Maier, J., Kaiser, P.D., Bodenhöfer, M., Schmees, C., Vojnovic, I. et al. (2018). A peptide tag-specific nanobody enables high-quality labeling for dSTORM imaging. *Nature Communications*, 9, 930.

Vogelsang, J., Cordes, T., Forthmann, C., Steinhauer, C., Tinnefeld, P. (2009). Controlling the fluorescence of ordinary oxazine dyes for single-molecule switching and superresolution microscopy. *Proceedings of the National Academy of Sciences of the USA*, 106, 8107–8112.

Wombacher, R., Heidbreder, M., van de Linde, S., Sheetz, M.P., Heilemann, M., Cornish, V.W., Sauer, M. (2010). Live-cell super-resolution imaging with trimethoprim conjugates. *Nature Methods*, 7, 717–719.

Woodman, R., Yeh, J.T.-H., Laurenson, S., Ko Ferrigno, P. (2005). Design and validation of a neutral protein scaffold for the presentation of peptide aptamers. *Journal of Molecular Biology*, 352, 1118–1133.

Xue, L., Karpenko, I.A., Hiblot, J., Johnsson, K. (2015). Imaging and manipulating proteins in live cells through covalent labeling. *Nature Chemical Biology*, 11, 917–923.

4

Quantitative FRAP and FCS

Cyril FAVARD

Infectious Disease Research Institute of Montpellier, IRIM, CNRS, France

4.1. Life is motion

Molecular dynamics in biological systems play a key role in the understanding of life events. Biological motions occur in a very large range of time and space scales. Several different approaches allow quantitative measurements of these motions at the molecular space and time scale (i.e. from nm to mm and from ns to s), each of them covering different areas of this large spatio-temporal range (see Figure 4.1). However, in the last 50 years, advances in optical microscopy and associated technologies (laser, detectors, electronics and computers) have made it possible to monitor these motions directly under the microscope within living samples. Moreover, the revolution of fluorescent proteins in the early 1990s (see Chapter 3) has allowed researchers to generate fluorescent chimeric proteins in living cells in order to stoichiometrically (1 protein for 1 fluorescent protein) follow their localization and their dynamics.

To monitor these molecular dynamics, two main types of methods can be differentiated: motion can be either analyzed for each molecule using single particle tracking (see Chapter 5) or motion can be characterized by monitoring the average dynamics of an ensemble of (few) molecules. Fluorescence Recovery After Photobleaching (FRAP) and Fluorescence Correlation Spectroscopy (FCS) belong to this second category. Both were designed and developed in the lab of W.W. Webb at Cornell University. However, they have a fundamental difference in their principle.

For a color version of all figures in this chapter, see www.iste.co.uk/sibarita/photonic.zip.

Photonic Imaging for Biology,
coordinated by Jean-Baptiste SIBARITA. © ISTE Ltd 2025.

FCS, the first technique to be released, beginning of the 1970s (Magde et al. 1972, 1974; Elson and Magde 1974), was originally developed to monitor chemical reaction kinetics at equilibrium. On the contrary, FRAP, released four years later (Axelrod et al. 1976), is based on equilibrium perturbation and relaxation dynamics monitoring. Mainly restricted to biophysics laboratories till the end of the 20th century, these two techniques, taking advantage of the fluorescent proteins revolution, started to appear in biology laboratories by the beginning of 2000s.

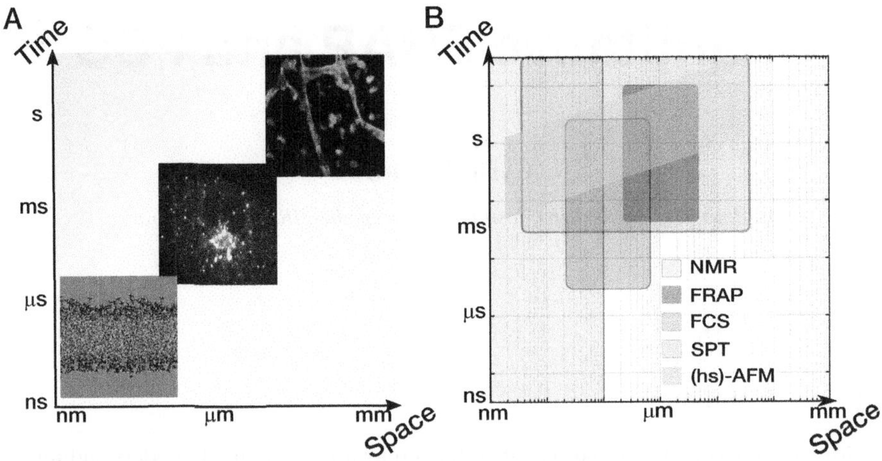

Figure 4.1. *Time and space scales of biological dynamic processes. A – Illustration of the temporal scale of different motions from molecules to tissues. B – Spatio-temporal range covered by different techniques allowing quantification of dynamics in biological samples*

FRAP is undoubtedly the most widely used technique in biology laboratories. It takes its advantage from its intrinsic experimental simplicity and immediate readout. However, as a counterpart, it is complex to quantitatively analyze FRAP data. FCS seems a more advanced technique as its readout is not immediate and necessitates a set of mathematical transformations of the raw signal, but surprisingly its quantitative analysis is much simpler.

NOTE.– *Molecular motions in cells:* numerous types of molecular motions and/or kinetics can be distinguished. Because the aim of this chapter is neither to describe all the possible situations nor to give an exhaustive physical description of these motions/kinetics, each of the two techniques (FRAP and FCS) will have a dedicated section giving analytical solutions found in the scientific literature for the different possibilities. These are detailed in sections 4.2.3 and 4.3.3.

This chapter is divided into two main parts, namely FRAP and FCS. In each part, we will first explain the basic concepts of each technique and then go into the details of the underlying physical principles and the quantitative analysis that could be obtained for the two techniques for different types of kinetics. Finally, we will illustrate recent advances in these techniques by giving examples of new techniques derived from these two original ones.

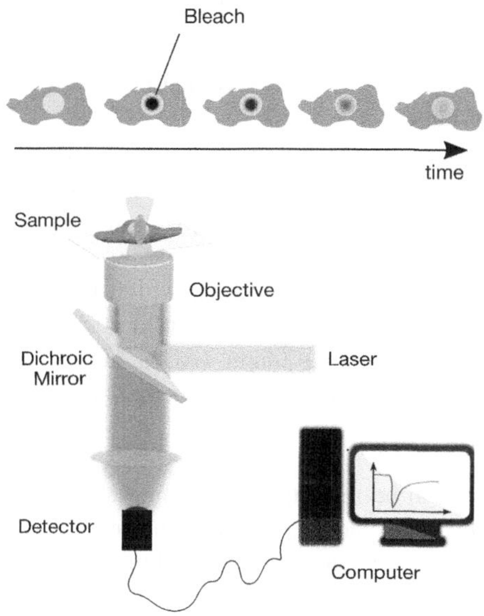

Figure 4.2. *A typical FRAP experiment. A laser is focused on the sample at the position where the molecule's dynamic is to be monitored. After a given time (prebleach), a high power of laser is delivered onto the sample (bleach); the fluorescence recovery is then monitored at the same position*

4.2. FRAP

Although it was designed after FCS, FRAP has taken over and become the most commonly used technique over the last 25 years in cell biology labs. FRAP exploits the phenomenon of photobleaching, by selectively and irreversibly "turning off" fluorescent molecules within a region of interest and analyzing the recovery of fluorescence within the same region. Its technical simplicity makes it nowadays routinely used in numerous biological research labs. Paradoxically, while being the simplest technique, it is often the most complicated to interpret properly.

4.2.1. *Basics of FRAP*

FRAP relies on the property of any fluorophore to be irreversibly photobleached (i.e. to experience a photochemical reaction) upon exposure to light.

Figure 4.2 shows a typical FRAP experiment under the microscope. The light source is focused on an area/volume of the cell. This FRAP experiment can be basically divided into three steps:

1) The pre-bleach step is a time-lapse experiment that will be used as a control prior to bleaching, where fluorescence of this area is recorded on the detector with a low-level illumination.

2) The bleaching step where the area/volume to be photobleached is flashed with a short and strong light pulse.

3) The recovery step, a time-lapse experiment usually (much) longer than the pre-bleach one, monitoring onto the detector the fluorescence recovery over time within the same photobleached area/volume, with a low-level illumination.

Therefore, any microscope equipped with a high-intensity illumination source (such as a laser), a fast illumination modulation system (such as an acousto-optic modulator or acousto-optic tunable filters (AOM/AOTF)) and a sensitive point detector or bidimensional detector (such as a camera) can perform FRAP experiments (Figure 4.2).

Different modalities can be imagined, for example, wide-field microscopes with a point source for bleaching and a classical illumination (lamp/LED) to monitor recovery in a 2D detector, laser scanning confocal microscopes with a point detector and an AOM that will allow a fast and accurate switch from a low-intensity illumination to a high-intensity illumination in less than a μs. A specific illumination profile can also be designed, for example, illumination by a fringe pattern has long been used to perform FRAP experiments. Finally, the use of scanners or spatial light modulators allows the generation of almost every possible illumination geometry.

An oversimplified analysis of FRAP experiments is based on a descriptive approach comparing the $t_{1/2}$ values, i.e. the time value when the recovery reaches 50% of the final fluorescence signal (Figure 4.3), without a priori knowledge of the motion process in different biological conditions.

This approach is often used in biology labs. An empirical parameter that is both qualitative and quantitative can be defined here, namely the mobile fraction, which is defined as the proportion of molecules that have left (or entered) the photobleached area. It is calculated from the $F(t = 0)$, Fo and F_∞ value as $M = \frac{F_\infty - F(t=0)}{Fo - F(t=0)}$ and varies from 0 to 1 (Figure 4.3). By definition, the immobile fraction is $1 - M$. This is often used to compare the free mobility of molecules, although it is important to recall here that this parameter has no absolute significance since it depends on the duration of the recovery monitoring and therefore cannot be easily compared from one study to another.

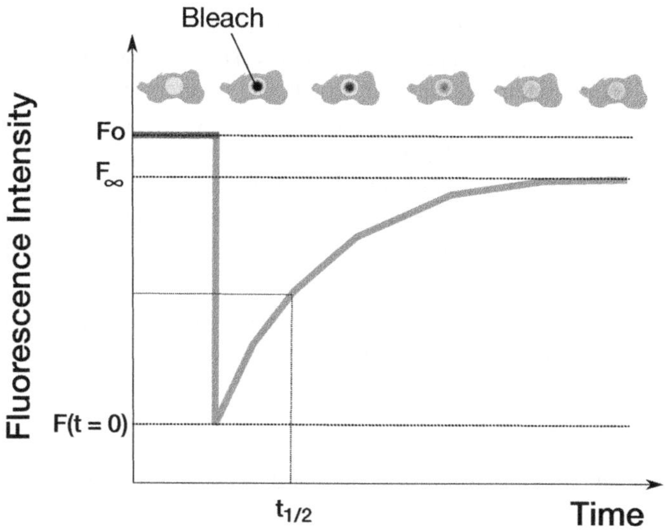

Figure 4.3. *A typical fluorescence recovery curve. The fluorescence intensity is monitored at the bleach position with time. Before the bleach, the fluorescence value is measured at F_0, immediately after the bleach, the value decreases down to $F(t = 0)$ and is then recovering at either F_0 if the mobility of the molecule is not restricted or at an F_∞ value if the mobility is not full (M < 1, see text for explanation)*

4.2.2. *Quantitative principles of single spot FRAP*

4.2.2.1. *Geometry of the photobleached area*

In the microscope, the fluorescent signal reaching the detector depends on several parameters. In a general approach (that will also be taken in section 4.3.2), this signal depends on the instrument and the fluorescent molecule. The intrinsic molecular fluorescence mainly depends on the molecular cross-section σ (equivalent to absorption coefficient) and the fluorescence quantum yield ϕ. The instrument parameters are the spatial distribution of the excitation intensity $I_{exc}(\vec{r})$, the collection efficiency, also varying with the position $D(\vec{r})$, other parameters such as detector yield, transmission efficiency, wavelength collection efficiency and so on that we can group into a single parameter that we will name κ. All this combined make it possible to write the following:

$$I = \kappa\, \sigma\, \phi \int I_{exc}(\vec{r})\; D(\vec{r})\; C(\vec{r})\; d^3\vec{r}, \qquad [4.1]$$

with $C(\vec{r})$ being the concentration of the fluorophore.

In FRAP, the photodestruction geometry is a key factor of the experiment, as the fluorescence recovery will depend on it. In recent microscopes such as laser (fast) scanner microscopes or even spatial light modulators, any type of bleaching geometries can be generated nowadays. However, whereas this could be of interest to bleach dedicated cellular structures, it often leads to a complex or non-existing analytical expression of the fluorescence recovery, making the analysis not quantitative and strongly dependent on this bleaching geometry (Blumenthal et al. 2015). Originally, two main geometries were used to perform FRAP, mainly depending on the origin of the illumination. In the following, we will mainly use these two simple illumination profiles to derive an analytical quantitative approach to fluorescence recovery:

1) Gaussian illumination. In the case of a laser beam focused on the sample through microscope objective, illumination can be written as:

$$I(\vec{r}) = I_0.e^{\frac{-2\vec{r}^2}{w^2}} \quad [4.2]$$

2) Circular illumination. A uniform circular disk of radius w can be defined as:

$$I(\vec{r}) = \begin{cases} I_0 & \vec{r} \leq w \\ 0 & \vec{r} > w \end{cases} \quad [4.3]$$

Solutions for more complex geometries have been derived from these two models, either in the case of a bleached 2D-disk with a radius much bigger than the waist of the laser in scanning confocal microscopes (Braeckmans et al. 2003) or in the case of a spot laser bleaching in 3D for with high numerical objectives in a confocal configuration (Blonk et al. 1993). On the opposite, other groups have designed simple geometries, such as strips, to obtain simple expressions for the recovery fit.

4.2.2.2. *Photobleaching*

Photobleaching occurs when a fluorophore M, brought to an excited state after light absorption, chemically reacts with another chemical species X (such as oxygen, for example) present in the sample, this reaction generating a new species MX that has different photophysical properties and is unable to fluoresce again. It is often approximated as a first-order process (see equation [4.5]). The reaction kinetics (equation [4.5]) show that the photobleaching efficiency is proportional to the laser intensity ($I(r)$) and to the exposure time (t), meaning that a short and strong light pulse will bleach as many molecules as a long and weak light pulse:

$$M + h\nu + X \xrightarrow{k_b} MX \quad [4.4]$$

$$\frac{dM_b}{dt} = -k_b I(\vec{r}) M \Rightarrow M_b = M_0 e^{-k_b I(\vec{r}) t} \quad [4.5]$$

Rigorously, the fluorescence recovery with time is always a convolution of the photobleaching kinetics and recovery kinetics and should be analyzed as such;

however, short and strong pulses of light can be considered as temporal Dirac functions, considerably simplifying the analytical expressions of the fluorescence recovery. In practice, this can be assumed if the bleaching time is much shorter than the characteristic recovery time (i.e. at least an order of magnitude shorter). If it is not the case, then complex behavior must be analyzed in detail (Waharte et al. 2005).

Immediately after the end of the photobleaching process, the concentration of the fluorophore can then be defined by combining equations [4.5] and [4.2] or [4.3] leading to:

$$c_M(\vec{r},0) = c_M(0)e^{-k_b.I_0.t.e^{\frac{-2\vec{r}^2}{w^2}}} \qquad [4.6]$$

in the case of a Gaussian spot illumination, or:

$$c_M(\vec{r},0) = \begin{cases} c_M(0)e^{-k_b.I_0.t} & \vec{r} \leq w \\ c_M(0) & \vec{r} > w \end{cases} \qquad [4.7]$$

in the case of a circular bleach.

Finally, the fluorescence observed at the end of the bleaching pulse, i.e. at time $t = 0$ of the recovery, will be given by combining equations [4.1] and [4.6] or [4.7] and defining a bleaching efficiency as $K = k_b I_0 t$ resulting in:

$$F_M(0) = I_0.c_M(0)\frac{1 - e^{-K}}{K} \qquad [4.8]$$

in the case of a Gaussian spot illumination, or:

$$F_M(0) = I_0.c_M(0)e^{-K} \qquad [4.9]$$

in the case of a circular bleach.

The recovery can now be determined (or not) analytically for different types of motions.

4.2.3. *Analytical expression for principal origins of fluorescence recoveries*

Hereafter are given analytical expressions for spot FRAP under Gaussian and (when available) circular profiles of photobleaching. Many models have been designed for more complex geometries, which will not be detailed in this chapter; in particular, in the case of imaging FRAP mentioned later in section 4.2.5. Contrary to FCS, in FRAP, the analytical solutions are rather difficult to handle numerically.

Free diffusion

This is the first model explored in the original paper of Axelrod et al. (1976). Solutions are given for both Gaussian and circular profiles, this later being revisited by Soumpasis (1983). In the case of Gaussian illumination, the analytical solution of recovery for a fully mobile molecule (no immobile fraction) is written as:

$$F_M(t) = I_0.c_M(0)\sum_{n=1}^{\infty}\frac{(-K)^n}{n!}\frac{1}{1+n(1+2\frac{t}{\tau_D})} \quad [4.10]$$

To express the solution for a circular profile, the recovered fluorescence $F_M(t)$ is first rewritten in terms of the fractional recovery $f(t)$, defined as:

$$f(t) = \frac{F_M(t) - F_M(0)}{F_M(\infty) - F_M(0)} \quad [4.11]$$

leading to a simple expression:

$$f(t) = e^{\frac{-2\tau_D}{t}}\{I_0\frac{2\tau_D}{t} + I_2\frac{2\tau_D}{t}\} \quad [4.12]$$

For a scanning confocal microscope, the problem is more complex. Many attempts have been made to simplify it. A nice summary of simplified expressions, compared to spot FRAP, in the case of pure diffusion molecular mobility can be found in Kang et al. (2012).

Flow

If the molecular motion is flow, the Gaussian illumination solution is then given by:

$$F_M(t) = I_0.c_M(0)\sum_{n=1}^{\infty}\frac{(-K)^n}{(n+1)!}e^{\frac{-2n}{n+1}(\frac{t}{\tau_F})^2} \quad [4.13]$$

The solution for the circular profile being quite simple:

$$f(t) = (2/\pi)sin^{-1}(t/2\tau_F) + (t/\tau_F)(1 - t^2/4\tau_F^2)^{1/2} \quad [4.14]$$

with $\tau_F = w/V_0$, V_0 being the flow speed.

Anomalous (sub)-diffusion

Twenty years later, the anomalous (sub)-diffusion problem was addressed by the team of W.W. Webb (Feder et al. 1996), where the FRAP technique was initially developed. In its simplest expression, it only introduces the anomalous exponent of the non-Brownian diffusion in the (Gaussian) free diffusion model:

$$F_M(t) = I_0.c_M(0)\sum_{n=1}^{\infty}\frac{(-K)^n}{n!}\frac{1}{1+n(1+2(\frac{t}{\tau_D})^{\alpha})} \quad [4.15]$$

Although convenient, this formulation is not strictly rigorous as it only deals with one of the numerous possible sources of anomalous diffusion. Moreover, in their original paper, this approach is complemented by the use of an immobile fraction to correctly fit the experimental data, which is theoretically questionable. This is the reason why many different studies have been conducted over the last 20 years on FRAP and anomalous diffusion. It is beyond the scope of this chapter to go into detail on this. However, the reader should keep in mind that the apparent anomaly of the experimental diffusion ($\alpha \neq 1$) could be observed even for motions that are not anomalous.

Reaction–diffusion

As in the case of anomalous diffusion, the FRAP expression for reaction–diffusion kinetics was analytically solved later on despite the fact that the reaction–diffusion case seems the most relevant to many, so-called "specific", biological processes (such as protein binding to DNA or lipid membrane or small molecule binding to a receptor, for example).

The equilibrium in this case can be written as: $A + B \underset{k_{off}}{\overset{k_{on}}{\rightleftharpoons}} AB$, where B is considered to diffuse much slower ($D_B = D_{AB} \to 0$) than A which has an infinite reservoir $[A] >> [B]$. Therefore, a pseudo-kinetic constant has to be defined as $k^*_{on} = k_{on}[B]$. The equilibrium can be rewritten as: $A_F \underset{k_{off}}{\overset{k^*_{on}}{\rightleftharpoons}} A_B$, where A_F is the unbound proportion of A and A_B is the bound one.

The concentration kinetics of each species then obeys the following set of equations:

$$\begin{cases} \frac{\partial A_B}{\partial t} = k^*_{on} A_F - k_{off} A_B \\ \frac{\partial A_F}{\partial t} = -D_{A_F} \nabla^2 A_F - k^*_{on} A_F + k_{off} A_B \end{cases} \qquad [4.16]$$

At the end of the recovery, the fluorescence intensity reaches a plateau that corresponds to the equilibrium and is generally normalized to 1. Therefore, at equilibrium, $A_B + A_F = 1$, and therefore, the equilibrium values of A_B and A_F can be reformulated as a function of k^*_{on} and k_{off} as:

$$\begin{cases} A_F(eq) = \frac{k_{off}}{k^*_{on} + k_{off}} \\ A_B(eq) = \frac{k^*_{on}}{k^*_{on} + k_{off}} \end{cases} \qquad [4.17]$$

This set of equations theoretically permits the determination of the kinetics constants k^*_{on} and k_{off}.

In the case of full reaction–diffusion kinetics, the analytical expression of the FRAP recovery curve only exists in the Laplace space and therefore necessitates a numerical inversion, which is beyond the scope of this chapter. The reader can refer to Sprague et al. (2004) for detailed calculations. However, in this full model, three regimes can be distinguished, which lead to a strong simplification of the analytical expression:

1) Pure diffusion regime, where reaction is not present and leads to the solutions expressed here (see equations [4.10] and [4.12]).

2) Effective diffusion regime, occurring when the reaction process is much faster than the diffusion process and leading to an apparent diffusion coefficient. In this case, the apparent diffusion coefficient is written as $D_{ap} = \frac{D_{A_F}}{1+\frac{k^*_{on}}{k_{off}}}$.

3) Reaction dominant is when the diffusion is much faster than the reaction kinetics, leading to the following expression:

$$f(t) = 1 - A_B(eq)e^{-k_{off}t} \quad [4.18]$$

In this case, combining equations [4.18] and [4.17] allows the determination of k^*_{on} and $A_F(eq)$, solving the full set of the reaction parameters.

4.2.4. *Variable radii FRAP*

FRAP experiments were originally developed using point detection and point excitation/bleaching, leading to a temporal information of the dynamics at one space scale/position. Due to the complexity of the above-described analytical model, and due to the often poor signal to noise recovery curve obtained experimentally, it is clear that retrieving exact kinetic values necessitates large statistics and that it is tricky to determine which kinetic model fits the experimental recovery the best. To circumvent this problem, multiple space scale FRAP was developed, using variable radii FRAP experiments, which consist of analyzing the changes in the recovery time or mobile fraction as a function of the photobleached area (Yechiel and Edidin 1987; Salomé et al. 1998).

Theoretically, recovery time can be plotted as a function of the size of the radius photobleached, authorizing direct distinction between the different possibilities of molecular motions ((anomalous) diffusion, flow, exchange, etc.) (Favard 2018) as this will be done in single particle tracking. In the case of diffusion in heterogeneous environments such as a membrane with lipid domains, the slope of the linear relationship between the mobile fraction and the inverse of the photobleached radius has been shown to give access to the average size of the domains (Salomé et al. 1998) (see Figure 4.4). This approach can therefore be used to quantitatively monitor the spatial heterogeneity sensed by the diffusing molecule.

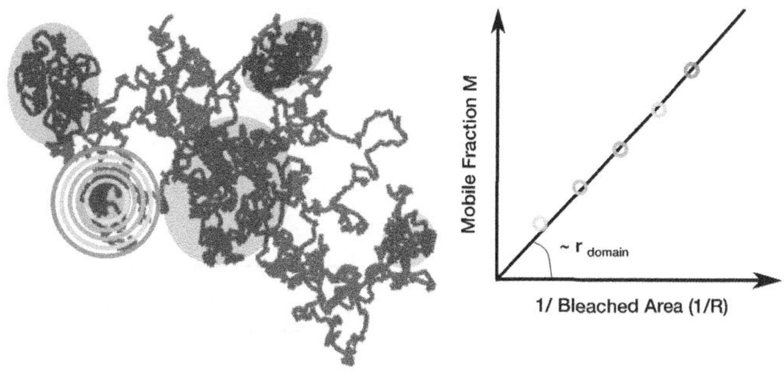

Figure 4.4. *Principle of vr-FRAP. FRAP is performed at different radii into the sample, and either the recovery time or the mobile fraction, M, can be plotted as a function of the bleached radius*

4.2.5. *Quantitative imaging FRAP*

This spatio-temporal approach is now possible thanks to the recent advances in scanners, detectors and different optical components, enabling imaging FRAP experiments. Imaging FRAP is based on rapidly photobleaching an area with an arbitrary shape and immediately imaging that fluorescence recovery in the photobleached area. This improvement of the technique has many advantages and opens infinite possibilities. The most trivial is the capacity to correct the fluorescence recovery obtained in the photobleached area by additional regions either located outside the cell (to remove background signal) or inside the cell (to correct for additional bleaching occurring during the recovery acquisition). These two corrections generally allow better determination of characteristic recovery time as well as mobile fraction (see Figure 4.5a). Imaging also makes it possible to simultaneously perform multiple measurements, generating much higher statistics or authorizing direct comparisons of kinetics at different locations of the same cell to test the spatial heterogeneity (Figure 4.5b). Finally, and importantly, quantitative imaging FRAP allows direct determination of the nature of the motion thanks to spatio-temporal analysis of the bleached spot refilling. Since the spatial dependency of diffusion or flow processes will induce either a widening or a displacement of the center of the bleached spot while binding/unbinding will result in an apparent refilling of the bleached spot without widening (see Figure 4.5c), these different processes can theoretically be easily discriminated.

Quantitative imaging FRAP also strongly facilitates the direct analysis of recovery curves by generating the recovery in the Fourier space, as the analytical model becomes much simpler and the recovery can be analyzed independently of the bleaching profile (Berk et al. 1993).

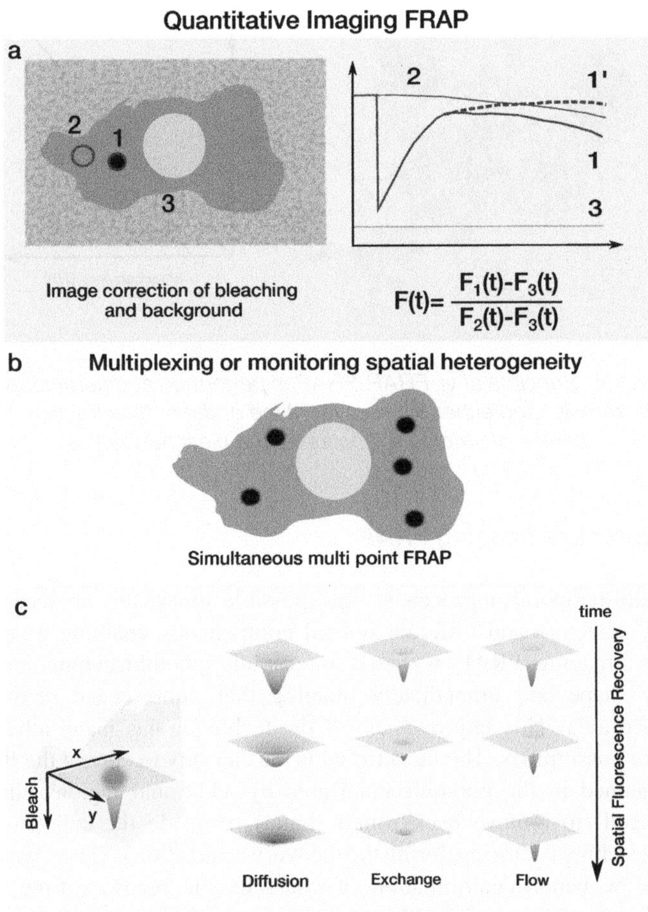

Figure 4.5. *Advantages of quantitative imaging FRAP. a – Imaging FRAP allows for direct correction of sample drift, photobleaching occurring while monitoring recovery and possible fluorescence background. b – Imaging FRAP allows for multiplexing, i.e. performing different bleach experiments simultaneously in the sample, either to accumulate data for better statistics or to monitor heterogeneity within the sample. c – Using a spatio-temporal analysis of the recovery, imaging FRAP can in theory allow direct discrimination of the kinetic generating the recovery (diffusion, exchange, flow, etc.)*

4.2.6. *Fluorescence loss in photobleaching*

Fluorescence loss in photobleaching (FLIP) consists of a continuous photobleaching of a chosen area (volume) of the cell introduced 25 years ago

(Cole et al. 1996). The bleached molecules diffuse into the immediate environment of the photobleached area (volume) while the one entering it are also bleached. This results in progressive extinction of the fluorescence in the cell that can be followed by simultaneous imaging during continuous photobleaching. This technique is not per se dedicated to be quantitative or qualitative in terms of molecular motions. It is mostly designed to visualize and understand the spatial continuity of compartments such as Golgi, Reticulum or Mitochondria within cells. It can also give insight into trafficking of (clusters of) molecules within cells. Of course, this conceptual approach can be extended to bigger objects such as embryos and organoids to monitor cell to cell exchange of matter. However, it must be kept in mind that continuous photobleaching might not be neutral to the cell/organism. Whereas an intense but very short (μs to ms) light pulse might have low photo-damaging consequences, continuous (s to min) although less powerful laser exposition could induce an irreversible chemical process in the cell, leading to cell death.

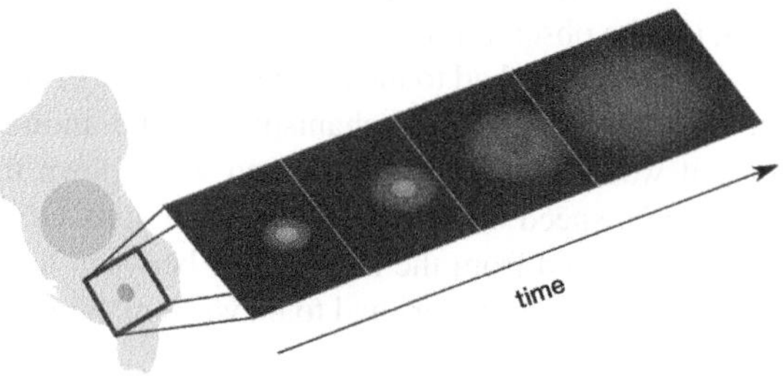

Figure 4.6. *Illustration of signal dilution of the photoconverted dye due to diffusion. As in the case of imaging FRAP, the spatio-temporal evolution of the fluorescence pattern is giving information about the nature of the mobility process and the homogeneity of the environment*

4.2.7. *Photoactivation/photoconversion*

In a very simple description, photoactivation or photoconversion could be considered as a "positive" photography of a photobleaching experiment. Instead of switching off the fluorophore in a well-defined area/volume, this experiment turns on the fluorescence (photoactivation) or swaps it to another wavelength (photoconversion) which, as a result, generates a positive signal in the area/volume enlightened (Patterson and Lippincott-Schwartz 2002). One of the biggest advantage is the ability to directly observe the spatio-temporal evolution of this generated signal. Therefore, as in the case of FRAP, the experimentalist can either focus on the change in fluorescence over time in the photoactivated area/volume, and fit the experimental curve with an analytical model to extract quantities of the different parameter tested, or can analyze the change in the spatio-temporal shape of the signal to retrieve more information (see Figure 4.6). Finally, imaging spatial propagation of newly generated fluorescence only without quantitative analyses can already provide information on the organization of the cell (spatial continuities or discontinuities of organelles, for example).

4.3. FCS

4.3.1. *Basic principle of correlation spectroscopy*

The origin of correlation spectroscopy is commonly attributed to the work of Svedberg and Inoue in 1911, counting colloidal particles in the microscope. In their paper they wrote:

> The colloidal solution to be examined was thus illuminated very briefly (0, 1, 2 sec.) at specific points in time equidistant from each other, which made it much easier to determine the number of particles. In order to get the zeros right, i.e. the observations in which no particle is in the field of view, contact pins were attached to the rotating disc, one of which closed the current to a small electric ring mechanism at each illumination. After some practice, it was possible to observe and write down the number of particles even at a speed of 39 exposures per minute. Of course, the eye could not be removed from the microscope between the individual determinations, so that the numbers had to be written "blind".

Figure 4.7 reproduces a series of individual counts of particles. When the number of observed particles is plotted versus time, we observe fluctuations of the number around an average value. If we now analyze a very high number of particles during a long time, we end up with a so-called "trace" of the particle number fluctuating around an average value. In their original work, Svedberg and Inoue plot the occurrence frequency of each number of particles counted at the same time $(n = 1,2,3,...)$ to determine the average number of particles that were present in their observation volume. By applying this method, they could measure the concentration of the colloid in their suspension.

This showed the potential of number fluctuations analysis to measure absolute concentrations of the observed molecule in samples. In addition, we can now temporally analyze the fluctuations by performing a temporal autocorrelation of the trace. For this, the trace is correlated with itself lagged at different time delays and normalized (the mathematical details of the process are given in section 4.3.2). This autocorrelation can then be plotted as a function of the time lags. Among the nice properties of this analysis, it is to be noted that the origin of the function is inversely proportional to the average number of particles present in the observation volume (Figure 4.7) and the abscissa corresponding to the half maximum of the function quantifies the average time the particle stays in the observation volume. This trace autocorrelation analysis gives an insight into the dynamics of the particle in its environment. This is the most commonly used nowadays in FCS. Modern correlation spectroscopy for biology is mainly arising from the work by Elson and Magde in the mid 1970s (Magde et al. 1972, 1974; Elson and Magde 1974).

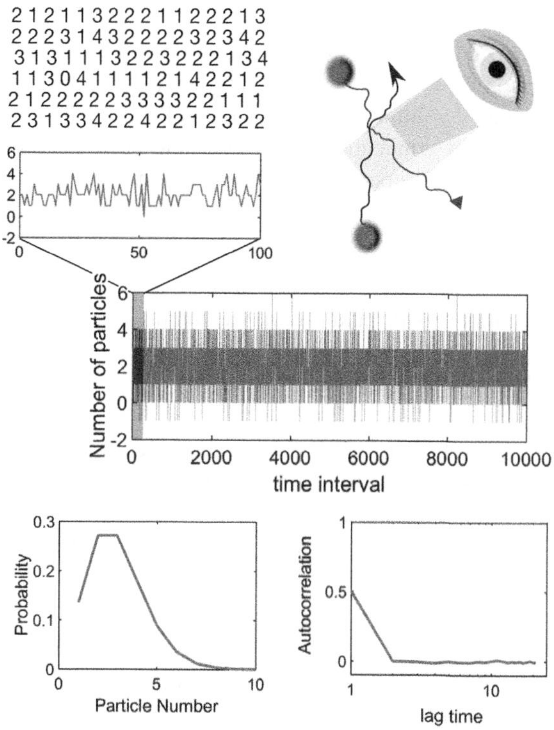

Figure 4.7. *Overview of the original experience from Svedberg and Inoue. The experimentalist is counting the average number of particles that are observable in the volume at definite time intervals. From these numbers, one can draw a graph of number fluctuations with time intervals*

4.3.2. *Mathematics of FCS*

As introduced before, FCS relies on temporal fluctuations of the fluorescence intensity $I(t)$. To easily monitor them, these temporal fluctuations need to be recorded in the smallest possible volume, justifying the use of confocal microscopes in laboratories. These fluctuations can occur in a solution or in a cell at a dynamic equilibrium, submitted to a linear fluorescence excitation regime. In this case, the fluorescence intensity $I(t)$ is proportional to the excitation intensity I_{exc} according to equation [4.19]:

$$I(t) = \kappa\, \sigma\, \phi \int I_{exc}(\vec{r})\; D(\vec{r})\; C(\vec{r},t)\; d^3\vec{r}, \qquad [4.19]$$

where κ is an instrumental parameter, σ and Φ are the molecular parameters (respectively absorption cross-section and quantum yield), $I_{exc}(\vec{r})$ is the excitation intensity in the illuminated volume and $D(\vec{r})$ is the collection efficiency in the object plane. Finally, $C(\vec{r},t)$ is the concentration at time t and position $\vec{r}$ of the fluorescent molecule. The recorded signal depends on the molecular detection efficiency $\zeta(\vec{r})$ that is defined as:

$$\zeta(\vec{r}) = I_{exc}(\vec{r})\; D(\vec{r}). \qquad [4.20]$$

By combining equations [4.19] and [4.20], we obtain:

$$I(t) = \kappa\, \sigma\, \Phi \int \zeta(\vec{r}) C(\vec{r},t)\; d^3\vec{r}, \qquad [4.21]$$

If we consider κ,σ,Φ to be constant over time and space, we now have two distinct contributions to the intensity fluctuations, one being instrumental ($\zeta(\vec{r})$), the second linked to the sample to analyze ($C(\vec{r},t)$). Now, if $I_{exc}(\vec{r})$ is constant over time, the intensity fluctuations $\delta I(t)$ can then be obtained by subtracting from $I(t)$ its own average value:

$$\delta I(t) = I(t) - \bar{I}, \qquad [4.22]$$

Considering therefore hereafter that these fluctuations $\delta I(t)$ are exclusively due to concentration variation $\delta C(\vec{r},t) = C(\vec{r},t) - \bar{C}$ occurring in the observed volume, these fluctuations can be rewritten as:

$$\delta I(t) = I(t) - \bar{I} = \kappa\, \sigma\, \Phi \int \zeta(\vec{r}) \delta C(\vec{r},t)\; d^3\vec{r}, \qquad [4.23]$$

4.3.2.1. *Analytical expression and properties of the temporal autocorrelation function*

Fluctuations in concentrations can be characterized using temporal correlation. We defined a normalized fluctuation function as $g^2(\tau)$:

$$g^2(\tau) = \frac{\langle I(t)\, I(t+\tau)\rangle}{\langle I(t)\rangle^2} = 1 + \frac{\langle \delta I(t)\; \delta I(t+\tau)\rangle}{\langle I(t)\rangle^2} = 1 + \frac{G(\tau)}{\langle I(t)\rangle^2}, \qquad [4.24]$$

This gives two extrema values. On the one side, when the fluctuations are totally decorrelated ($\tau \to \infty$), we obtain:

$$\lim_{\tau \to +\infty} g^{(2)}(\tau) = 1. \qquad [4.25]$$

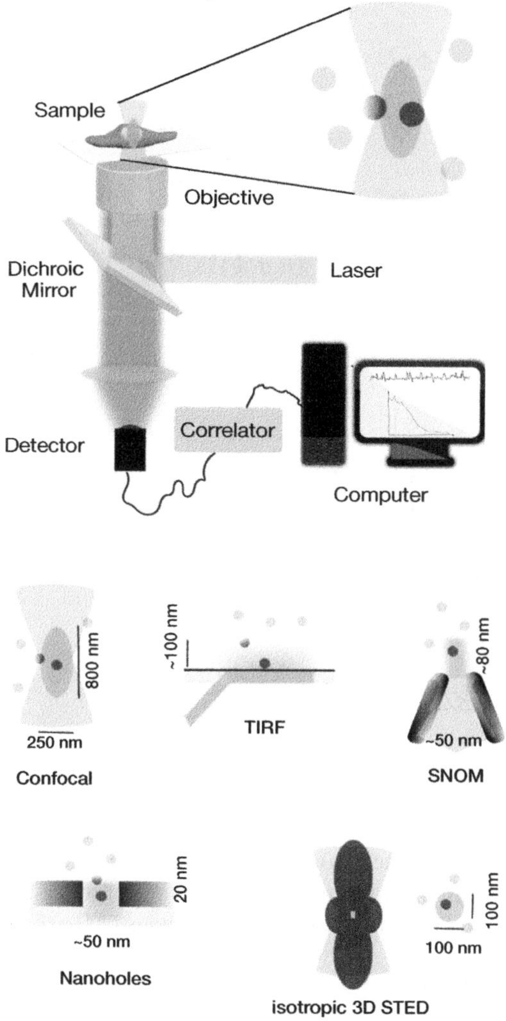

Figure 4.8. *Principle of an FCS set-up and its different optical configurations. In an FCS set-up, a laser is focused on the sample with a microscope, the fluorescence emitted is collected through the objective and focused on a highly sensitive detector coupled to a hardware correlator. The optical configuration of the microscope necessitates high spatial contrast that can be achieved not only by confocal or STED mode in far field, but also by TIRF, SNOM or nanoholes in near field excitation*

On the other side, when ($\tau \to 0$), we can obtain the average number of molecules in the volume. Since, for ergodic concentration fluctuations, we have $\langle \delta C^2 \rangle = \bar{C}$, we obtain:

$$\langle I(t) \rangle = \bar{I} = (\kappa\, \sigma\, \Phi)\, \bar{C} \int \zeta(\vec{r})\, d^3\vec{r}. \quad [4.26]$$

and thereafter:

$$g^{(2)}(0) = 1 + \frac{\langle \delta I(t)^2 \rangle}{\langle I(t) \rangle^2} = 1 + \frac{(\kappa\, \sigma\, \Phi)^2\, \bar{C} \int [\zeta(\vec{r})]^2\, d^3\vec{r}}{(\kappa\, \sigma\, \Phi)^2\, \bar{C}^2 \left[\int \zeta(\vec{r})\, d^3\vec{r}\right]^2} = 1 + \frac{1}{\bar{C}\, V_{eff}}, \quad [4.27]$$

with

$$V_{eff} = \frac{\left[\int \zeta(\vec{r})\, d^3\vec{r}\right]^2}{\int [\zeta(\vec{r})]^2\, d^3\vec{r}}. \quad [4.28]$$

The average number of molecules N in the detection volume is then given by $N = \bar{C} V_{eff}$ and equation [4.27] can be rewritten as:

$$g^{(2)}(0) = 1 + \frac{1}{N}. \quad [4.29]$$

4.3.3. *Analytical expression for different origins of concentration fluctuations*

A simple way to rewrite the autocorrelation function ($g^{(2)}(\tau)$) is to divide it into two parts, a constant term that reflects the average concentration, as in equation [4.29], and a second one that reflects mobility of the molecules ($M(\tau)$):

$$g^{(2)}(\tau) = g^{(2)}(0) M(\tau) \quad [4.30]$$

Different processes can account for the origin of temporal concentration fluctuations that have different biological significances. The most famous is free diffusion, also known as the Brownian motion. Hereafter, we only present a set of the most common analytical expressions for different cases without detailing the derivation of these expressions. If needed, the reader can refer to the FCS Expert website[1] and the different articles cited hereafter.

1. See www.fcsxpert.com.

Free diffusion

In the case of 3D free diffusion, we have:

$$M(\tau) = \frac{1}{1+\frac{t}{\tau_D}} \frac{1}{(1+\sigma^2\frac{t}{\tau_D})^{1/2}} \quad [4.31]$$

with σ being a shape factor taking into account the elongation of the Gaussian excitation profile defined as $\sigma = \frac{\omega_{xy}}{\omega_z}$. For 2D diffusion, the expression is:

$$M(\tau) = \frac{1}{1+\frac{t}{\tau_D}} \quad [4.32]$$

Flow

The flow condition can be split into two distinct models depending on the presence or absence of concomitant diffusion. For a unidirectional and non-lamellar flow without diffusion, we have (Magde et al. 1978):

$$M(\tau) = e^{-(\frac{\tau}{\tau_F})^2} \quad [4.33]$$

that become:

$$M(\tau) = \frac{1}{1+\frac{t}{\tau_D}} e^{-(\frac{\tau}{\tau_F})^2 \frac{1}{1+\frac{t}{\tau_D}}} \quad [4.34]$$

when 2D diffusion is also present. The above-described types of motion lead to nicely simple analytical expressions. However, they illustrate only very rare cases of biological molecule motions in their cellular/organism environment. Actually, most of the biologically relevant molecular motions occur in (highly) heterogeneous environments or correspond to diffusion–reaction processes. In both cases, the analytical expressions become more tricky.

Anomalous (sub)-diffusion

For (sub)-diffusion in a heterogeneous environment, 3D anomalous (sub)-diffusion analytical expression of $M(\tau)$ is always considered as being as simple as:

$$M(\tau) = \frac{1}{1+(\frac{t}{\tau_D})^\alpha} \frac{1}{(1+\sigma^2(\frac{t}{\tau_D})^\alpha)^{1/2}} \quad [4.35]$$

However, it must be kept in mind that, since the ergodicity condition is a prerequisite for FCS, equation [4.37] is not valid for anomalous diffusion process breaking this ergodicity such as certain sub-diffusive continuous time random walk (CTRW) processes, for example. Moreover, due to the divergence of the time-dependent diffusion coefficient on short time-scales, the α value obtained with equation [4.37] will always be lower than the real anomalous exponent one for $\alpha < 0.9$, leading to an overestimation of the sub-diffusion process.

Reaction–diffusion

Here, the same type of binding process is considered as in section 4.2.3.

This situation, however, does not lead to a single analytical expression. We have to distinguish several possible regimes depending on the characteristic times of each process (reaction $\propto k^*_{on}, k_{off}$ vs diffusion $\propto \tau_D$) (Michelman-Ribeiro et al. 2009). In the case where the reaction dominates towards free diffusion ($k^*_{on} >> k_{off}$), the solution is then given by:

$$M(\tau) = F_{eq} M_D(\tau) + C_{eq} e^{-k_{off} t} \qquad [4.36]$$

with

$$M_D(\tau) = \frac{1}{1 + \frac{t}{\tau_D}} \frac{1}{(1 + \sigma^2 \frac{t}{\tau_D})^{1/2}}$$

$$F_{eq} = \frac{k_{off}}{k_{off} + k^*_{on}} \quad C_{eq} = \frac{k^*_{on}}{k_{off} + k^*_{on}}$$

4.3.4. *Space and time correlation spectroscopy*

As previously stated for FRAP (see section 4.2.5), the determination and quantification of the correct process of molecular mobility mainly rely on the accuracy of the fit of the experimental data by the different analytical models exposed above, this accuracy being quantified using different types of statistical tests. However, experimental data in biology are often very noisy, and this approach rapidly becomes inaccurate. To overcome this issue, one solution is to perform spatio-temporal characterization of the molecular motion. For that purpose, several techniques derived from the original FCS approach have been developed over the last 20 years.

4.3.4.1. *Spot variation fluorescence correlation spectroscopy*

The spot variation fluorescence correlation spectroscopy (sv-FCS), as in the case of the variable radii FRAP (vr-FRAP), explores the variation of the decorrelation half-time ($\tau_{1/2}$) as a function of the square of the laser spot waist (area) in which fluctuations are monitored (Wawrezinieck et al. 2005). Plot of $\tau_{1/2} = f(\omega^2)$ gives an immediate insight into the nature of the motion as mean square displacement plots can be done in the single particle tracking experiments (see section 4.2.5). Simulations have been performed for various models that account for different deviations from 2D Brownian motion due to either lipid domains or obstruction by obstacles or by fences (Wawrezinieck et al. 2005) and, more recently, induced by solid–liquid phase separation (Favard et al. 2011). This approach has also been developed in super-resolution stimulated emission depletion (STED) microscopy (Eggeling et al. 2009) as well as more recently in imaging FCS (Veerapathiran and Wohland 2018).

4.3.4.2. *Line scanning fluorescence correlation spectroscopy*

One of the major problems with FCS in live biological samples is the many different temporal scales over which motion can happen. Low-frequency fluctuations can occur during signal recording at one position of the sample. For example, when monitoring diffusion of a lipid in the membrane, these low temporal frequency fluctuations can be due to global motion of the membrane occurring in living cells. This results in the appearance of correlations that could occur at different times, strongly perturbing the decorrelation due to the lipid motion. To overcome that issue, line scanning fluorescence correlation spectroscopy (ls-FCS) has been developed (Ries et al. 2009). In this technique, the laser is fast scanned on the sample and the intensity fluctuations are recorded along the different pixels of the scanned line. If global motion occurs during the line scan, it will displace the signal onto other pixels of the scanned line. Registration can then be made to get rid of these global motions. ls-FCS also offers many other advantages compared to point scanning: it decreases the photobleaching on the sample, it permits multiplexing (several points are scanned in parallel) and, more importantly, it has the ability to explore, in the same experiment, spatial heterogeneities of molecular motions in the sample.

This approach has been recently derived for super-resolution STED microscopy (Honigmann et al. 2014), making it possible, for example, to decipher lipid trapping during virus assembly in cells (Favard et al. 2019).

4.3.4.3. *Spatio-temporal image correlation spectroscopy-related techniques (TICS, STICS, RICS, lm-FCS)*

An alternative to line scanning fluorescence correlation spectroscopy (ls-FCS), which can be seen as a temporal correlation of different single points along one dimension, consists of the temporal correlation of two-dimensional images. Originally, this approach was developed with confocal images, as the Gaussian profile of the illumination spot makes it possible, as in the case of FCS, to easily compute and fit the autocorrelation curve, and comes from the Imaging Correlation Spectroscopy (ICS) technique developed by Petersen et al. (1993) in the late 1980s. ICS consists of the spatial correlation of images not subjected to any dynamic fluctuations, i.e. no change in the fluorescence intensity with time due to molecular motion. This could be seen as monitoring the changes in $G(0)$ with space, i.e. monitoring the change in molecule numbers in a given area. We can define a generalized spatio-temporal autocorrelation function as:

$$r(\xi, \eta, \tau) = \frac{< \delta i(x, y, t).\delta i(x + \xi, y + \eta, t + \tau) >}{< i(x, y, t) >_t . < i(x, y, t + \tau) >_{t+\tau}} \quad [4.37]$$

where τ is the temporal lag of the signal, and ξ and η are the spatial lags of the signal in, respectively, x- and y-directions.

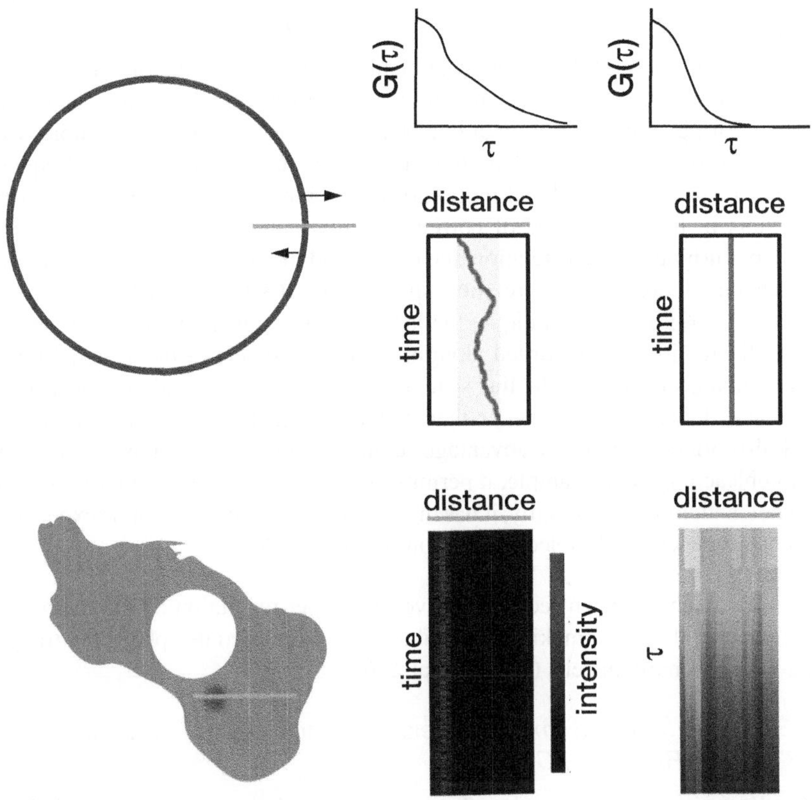

Figure 4.9. *Principle of line-scanning FCS. A line (or a circle) is scanned over the sample, and the intensity fluctuations with time are recorded at each position of the line. Signal analysis can then make it possible to correct for sample motion occurring during recording, in order to eliminate unwanted long-time decorrelation, or each time fluctuation at each point of the line scanned can be temporally autocorrelated in order to identify spatial heterogeneity in the sample*

For pure spatial correlation, this expression is simplified to:

$$r(\xi,\eta,0) = \frac{<\delta i(x,y,t).\delta i(x+\xi,y+\eta,t)>}{<i(x,y,t)>_t^2} \qquad [4.38]$$

and it can be shown that:

$$r(\xi,\eta,0) = g(0,0,0).e^{\frac{-\xi^2+\eta^2}{\omega^2}} + g_\infty \qquad [4.39]$$

where $g(0,0,0)$ is proportional to the molecule's cluster density present in the image.

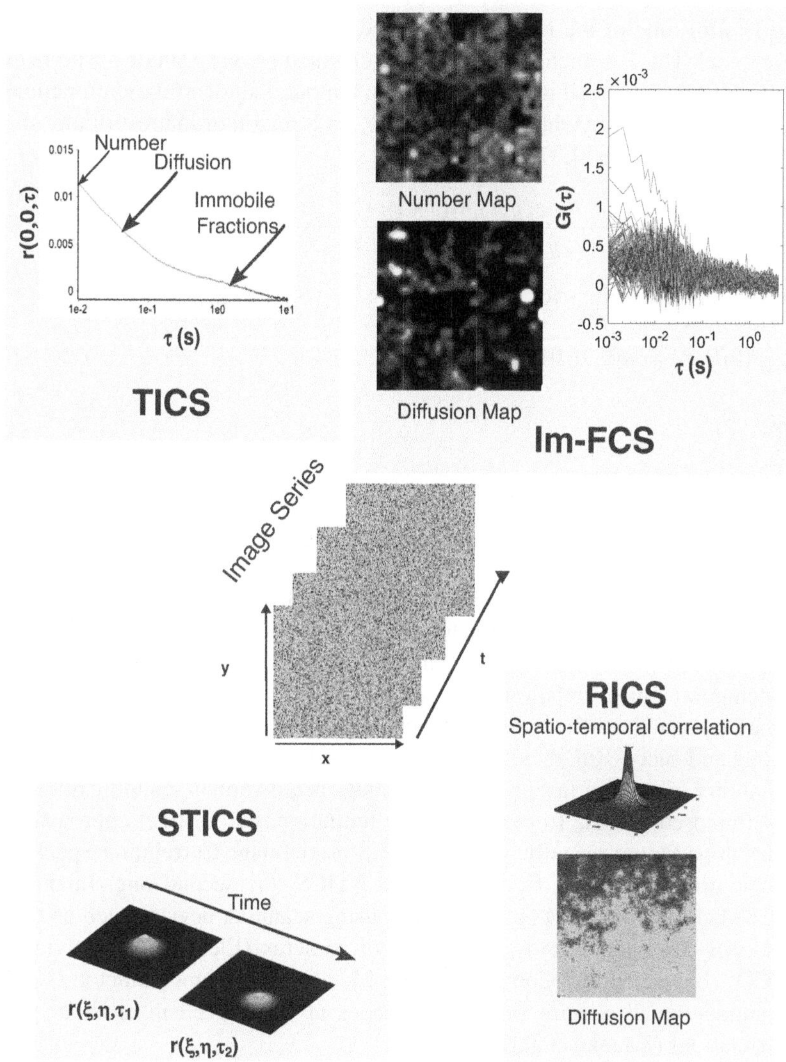

Figure 4.10. *Spatio-temporal correlation of images. The upper part of the figure depicts two modalities of temporal correlation of images, TICS and Im-FCS. The lower part represents two modalities of time evolution of spatial correlation of images, STICS and RICS. Each of these techniques has spatial and temporal distinct resolutions. Each address different ranges of molecular mobilities (from μs to ms: Im-FCS and RICS, from s to tens of s: TICS and STICS)*

Temporal image correlation spectroscopy (TICS) is a direct adaptation of this technique (Srivastava and Petersen 1996). To simplify, it consists of monitoring the evolution with time of the mean fluorescence of the image at a given (or averaged) position/pixel. This is somehow equivalent to classical FCS at a single position, except that now, thanks to the full image acquisition, temporal autocorrelation functions can be done a posteriori, anywhere in the sample, on a region of interest of any surface. Based on this, equation [4.37] now becomes:

$$r(0,0,\tau) = \frac{< \delta i(x,y,t).\delta i(x+,y+,t+\tau) >}{< i(x,y,t) >_t . < i(x,y,t+\tau) >_{t+\tau}} \quad [4.40]$$

and has different solutions for (2D) diffusion:

$$r(0,0,\tau) = g(0,0,0).(1+\frac{\tau}{\tau_D})^{-1} \quad [4.41]$$

or flow:

$$r(0,0,\tau) = g(0,0,0).e^{-(\frac{\tau}{\tau_f})^2} \quad [4.42]$$

As expected, both solutions nicely resemble to equations [4.32] and [4.33].

An adaptation of TICS is the spatio-temporal ICS, which, in comparison to TICS, makes it possible to directly estimate, in the image, the existence of directionality in the molecule's motions. In fact, changes in the shape and the position of the spatio-temporal autocorrelation function calculated in STICS give instantaneous information about the diffusion and flow processes (Hebert et al. 2005). Although very promising and successful, these two techniques suffer from a major drawback, which is their ability to monitor fast processes, as image acquisition in scanning microscopes is slow (hundred(s) ms). To overcome this limitation, two distinct approaches have been developed more recently: the first one, Raster Image Correlation Spectroscopy (RICS) is directly derived from TICS and STICS, the second one, Imaging-FCS (Im-FCS) is based on TICS but, instead of using scanning devices such as confocal microscopes, uses the improved capacity of cameras (Electron Multiplying CCD (EMCCD) and scientific Complementary Metal-Oxide-Semiconductor (sCMOS)) and computation to acquire full-frame images at high speed in order to perform autocorrelations (Kannan et al. 2006, 2007).

Im-FCS can be basically considered as a simultaneous multipoint temporal autocorrelation function, which allows the direct generation of correlogram maps as in the case of ls-FCS. The autocorrelation functions are the same as the one obtained in the case of single-point FCS and can be fitted according to the different models detailed in section 4.3.3. This is the nice part of this technique, which, however, is limited by the frequency rate of EMCCD or sCMOS full-frame chipset reading and the limited number of photons detected at high temporal frequency with this type of camera that are far less sensitive than classical avalanche photodiode (APD).

However, the flexibility of multidimensional detectors allows pixel binning to increase the number of photons while losing spatial resolution. Overall, a good experimental design needs to find the best compromise of these two parameters.

RICS has been developed for confocal scanning microscopy in order to extract mobility parameters directly from one to a few images (Digman et al. 2005). In the case of RICS, the spatial autocorrelation function $G_s(\xi,\eta)$ is defined as in equation [4.38] and now contains two terms $G_s(\xi,\eta) = G_\tau(\xi,\eta).S(\xi,\eta)$, the first one $G_\tau(\xi,\eta)$ being due to molecular motion, the second due to laser scanning $S(\xi,\eta)$. In both functions is found the mobility of the molecule of interest, which is usually considered as diffusion only to simplify the equations. The first term is defined as:

$$G_\tau(\xi,\eta) = \frac{\gamma}{N}\left(1+\frac{4D(\tau_p\xi+\tau_l\eta)}{w_0^2}\right)^{-1}\left((1+\frac{4D(\tau_p\xi+\tau_l\eta)}{w_z^2}\right)^{-1/2} \quad [4.43]$$

and is reminiscent of the case of 3D diffusion in single point FCS (equation [4.31]), with γ being an instrumental correction factor and τ_p and τ_l being the time spent, respectively, on a pixel or to scan a line. The correlation due to scanning is defined as:

$$S(\xi,\eta) = exp\left(-\frac{\frac{1}{2}[(\frac{2\xi\delta r}{w_0})^2 + (\frac{2\eta\delta r}{w_0})^2]}{(1+\frac{4D(\tau_p\xi+\tau_l\eta)}{w_0^2})}\right) \quad [4.44]$$

with δr being the size of the pixel.

	D ($m^2.s^{-1}$)	Time to adjacent point	Method
Recovered by spatial correlation	100	0.1 ms	Pixel/line/raster scanning
	10	1 ms	Pixel/line/raster scanning
	1	10 ms	Line/raster scanning
	0.1	100 ms	Line/raster scanning
	0.01	1,000 ms	Impractical
	0.001	10,000 ms	Impractical
	D ($m^2.s^{-1}$)	**Time at same point**	**Method**
Recovered by temporal correlation	100	< 0.01 ms	Point FCS
	10	< 0.1 ms	Point and line scanning FCS
	1	< 1 ms	Line/point and line scanning FCS
	0.1	< 10 ms	Line/point/line scanning FCS, Im FCS
	0.01	< 100 ms	Line scanning, TICS, ImFCS
	0.001	< 1,000 ms	TICS

Table 4.1. *Order of magnitude of diffusion coefficients that can be determined by spatial and time correlation using laser scanning systems (adapted from Digman et al. (2005))*

From this analysis, it is possible to retrieve the diffusion coefficient of the molecule within the sample. However, as with the other spatio-temporal correlation techniques, this necessitates a fine tuning of the pixel and/or the line scan duration to be in phase with the typical time of pixel/line occupancy by the molecule.

This is shown in Table 4.1, where the best autocorrelation method for each time-scale is given for both spatial correlation and temporal correlation.

4.4. Conclusion

As described in this chapter, many techniques have been derived from the original FRAP and FCS during the last half a century. Both FRAP and FCS are now commonly used in biology labs (FRAP being the more frequently used technique). However, quantification of molecular motion using either technique still requires good knowledge of the underlying processes and the equivalent analytical models. For both FRAP and FCS, different pieces of software have been developed to perform quantitative analysis of the experiments. Finally, FRAP, FCS or some of the derived advanced techniques (such as spot-variation FCS, RICS) are now even accessible in commercial microscopes, which makes them simple, user-friendly and thus even more accessible to untrained users.

4.5. References

Axelrod, D., Koppel, D.E., Schlessinger, J., Elson, E., Webb, W.W. (1976). Mobility measurement by analysis of fluorescence photobleaching recovery kinetics. *Biophysical Journal*, 16, 1055–1069.

Berk, D., Yuan, F., Leunig, M., Jain, R. (1993). Fluorescence photobleaching with spatial Fourier analysis: Measurement of diffusion in light-scattering media. *Biophysical Journal*, 65(6), 2428–2436.

Blonk, J.C.G., Don, A., Van Aalst, H., Birmingham, J.J. (1993). Fluorescence photobleaching recovery in the confocal scanning light microscope. *Journal of Microscopy*, 169(3), 363–374.

Blumenthal, D., Goldstien, L., Edidin, M., Gheber, L.A. (2015). Universal approach to FRAP analysis of arbitrary bleaching patterns. *Scientific Reports*, 5(1), 11655.

Braeckmans, K., Peeters, L., Sanders, N.N., De Smedt, S.C., Demeester, J. (2003). Three-dimensional fluorescence recovery after photobleaching with the confocal scanning laser microscope. *Biophysical Journal*, 85, 2240–2252.

Cole, N.B., Smith, C.L., Sciaky, N., Terasaki, M., Edidin, M., Lippincott-Schwartz, J. (1996). Diffusional mobility of golgi proteins in membranes of living cells. *Science*, 273(5276), 797–801.

Digman, M.A., Brown, C.M., Sengupta, P., Wiseman, P.W., Horwitz, A.R., Gratton, E. (2005). Measuring fast dynamics in solutions and cells with a laser scanning microscope. *Biophysical Journal*, 89(2), 1317–1327.

Eggeling, C., Ringemann, C., Medda, R., Schwarzmann, G., Sandhoff, K., Polyakova, S., Belov, V.N., Hein, B., von Middendorff, C., Schoenle, A., Hell, S.W. (2009). Direct observation of the nanoscale dynamics of membrane lipids in a living cell. *Nature*, 457(7233), 1159–1162.

Elson, E.L. and Magde, D. (1974). Fluorescence correlation spectroscopy. I. Conceptual basis and theory. *Biopolymers*, 13(1), 1–27.

Favard, C. (2018). Numerical simulation and FRAP experiments show that the plasma membrane binding protein PH-EFA6 does not exhibit anomalous subdiffusion in cells. *Biomolecules*, 8(3), 90.

Favard, C., Wenger, J., Lenne, P.-F., Rigneault, H. (2011). FCS diffusion laws in two-phase lipid membranes: Determination of domain mean size by experiments and Monte Carlo simulations. *Biophysical Journal*, 100, 1242–1251.

Favard, C., Chojnacki, J., Merida, P., Yandrapalli, N., Mak, J., Eggeling, C., Muriaux, D. (2019). HIV-1 gag specifically restricts PI(4,5)p2 and cholesterol mobility in living cells creating a nanodomain platform for virus assembly. *Science Advances*, 5(10), eaaw8651.

Feder, T., Brust-Mascher, I., Slattery, J., Baird, B., Webb, W. (1996). Constrained diffusion or immobile fraction on cell surfaces: A new interpretation. *Biophysical Journal*, 70(6), 2767–2773.

Hebert, B., Costantino, S., Wiseman, P.W. (2005). Spatiotemporal image correlation spectroscopy (STICS) theory, verification, and application to protein velocity mapping in living CHO cells. *Biophysical Journal*, 88(5), 3601–3614.

Honigmann, A., Mueller, V., Ta, H., Schoenle, A., Sezgin, E., Hell, S.W., Eggeling, C. (2014). Scanning STED-FCS reveals spatiotemporal heterogeneity of lipid interaction in the plasma membrane of living cells. *Nature Communications*, 5(1), 5412.

Kang, M., Day, C.A., Kenworthy, A.K., DiBenedetto, E. (2012). Simplified equation to extract diffusion coefficients from confocal FRAP data. *Traffic*, 13(12), 1589–1600.

Kannan, B., Har, J.Y., Liu, P., Maruyama, I., Ding, J.L., Wohland, T. (2006). Electron multiplying charge-coupled device camera based fluorescence correlation spectroscopy. *Analytical Chemistry*, 78(10), 3444–3451.

Kannan, B., Guo, L., Sudhaharan, T., Ahmed, S., Maruyama, I., Wohland, T. (2007). Spatially resolved total internal reflection fluorescence correlation microscopy using an electron multiplying charge-coupled device camera. *Analytical Chemistry*, 79(12), 4463–4470.

Magde, D., Elson, E., Webb, W.W. (1972). Thermodynamic fluctuations in a reacting system – Measurement by fluorescence correlation spectroscopy. *Physical Review Letters*, 29(11), 705–708.

Magde, D., Elson, E.L., Webb, W.W. (1974). Fluorescence correlation spectroscopy. II. An experimental realization. *Biopolymers*, 13(1), 29–61.

Magde, D., Webb, W.W., Elson, E.L. (1978). Fluorescence correlation spectroscopy. III. Uniform translation and laminar flow. *Biopolymers*, 17(2), 361–376.

Michelman-Ribeiro A., Mazza, D., Rosales, T., Stasevich, T.J., Boukari, H., Rishi, V., Vinson, C., Knutson, J.R., McNally, J.G. (2009). Direct measurement of association and dissociation rates of DNA binding in live cells by fluorescence correlation spectroscopy. *Biophysical Journal*, 97(1), 337–346. doi: 10.1016/j.bpj.2009.04.027.

Patterson, G.H. and Lippincott-Schwartz, J. (2002). A photoactivatable GFP for selective photolabeling of proteins and cells. *Science*, 297(5588), 1873–1877.

Petersen, N., Höddelius, P., Wiseman, P., Seger, O., Magnusson, K. (1993). Quantitation of membrane receptor distributions by image correlation spectroscopy: Concept and application. *Biophysical Journal*, 65(3), 1135–1146.

Ries, J., Chiantia, S., Schwille, P. (2009). Accurate determination of membrane dynamics with line-scan FCS. *Biophysical Journal*, 96(5), 1999–2008.

Salomé, L., Cazeils, J.-L., Lopez, A., Tocanne, J.-F. (1998). Characterization of membrane domains by frap experiments at variable observation areas. *European Biophysics Journal*, 27(4), 391–402.

Soumpasis, D. (1983). Theoretical analysis of fluorescence photobleaching recovery experiments. *Biophysical Journal*, 41(1), 95–97.

Sprague, B.L., Pego, R.L., Stavreva, D.A., McNally, J.G. (2004). Analysis of binding reactions by fluorescence recovery after photobleaching. *Biophysical Journal*, 86, 3473–3495.

Srivastava, M. and Petersen, N.O. (1996). Image cross-correlation spectroscopy: A new experimental biophysical approach to measurement of slow diffusion of fluorescent molecules. *Methods in Cell Science*, 18(1), 47–54.

Svedberg, T. and Inouye, K. (1911). *Zeitschrift für Chemie und Industrie der Kolloide (Colloïdologie)*. Steinkopff and Springer, Dresden.

Veerapathiran, S. and Wohland, T. (2018). The imaging FCS diffusion law in the presence of multiple diffusive modes. *Methods*, 140–141, 140–150.

Waharte, F., Brown, C.M., Coscoy, S., Coudrier, E., Amblard, F. (2005). A two-photon frap analysis of the cytoskeleton dynamics in the microvilli of intestinal cells. *Biophysical Journal*, 88, 1467–1478.

Wawrezinieck, L., Rigneault, H., Marguet, D., Lenne, P. (2005). Fluorescence correlation spectroscopy diffusion laws to probe the submicron cell membrane organization. *Biophysics Journal*, 89(6), 4029–4042.

Yechiel, E. and Edidin, M. (1987). Micrometer-scale domains in fibroblast plasma membranes. *Journal of Cell Biology*, 105(2), 755–760.

5

Single-Particle Tracking for Nanoscale Dynamics of Biological Samples

Antony LEE[1,2] and Laurent COGNET[1]

[1] *Laboratoire Photonique, Numérique et Nanosciences, Institut d'Optique Graduate School, UMR 5298, CNRS, University of Bordeaux, France*

[2] *Physique des Cellules et Cancer, Institut Curie, UMR 168, CNRS, Université PSL, Sorbonne Université, France*

5.1. Introduction

Understanding the complexity of living biological samples requires elucidating the dynamic organization of their constitutive components down to the nanoscale. Molecules in living cells or tissues move in a wide variety of manners over different spatio-temporal scales, making the study of their movements challenging. This is, for example, the case of membranes, which are highly complex environments due to the co-existence of a variety of lipids and proteins in dynamic and compartmentalized spatial distributions. There, molecules display various diffusion modes, which can relate to spatially localized inter-molecular interactions, and impact molecular functions.

Single-particle tracking methods, based on our capacity to resolve the movement of molecules at the single molecule level, has allowed the elucidation of many

---

For a color version of all figures in this chapter, see www.iste.co.uk/sibarita/photonic.zip.

Photonic Imaging for Biology,

molecular behaviors. Because fluorescence microscopy is well adapted to image live samples with high specificity, the study of dynamical biological processes by single-particle tracking is frequently based on our ability to efficiently detect the fluorescence of single nano-objects. In this chapter, we will describe how such experiments can yield nanoscale localization and trajectory information of the nano-probes, and how to infer biophysical parameters, such as the diffusion coefficient, the anomalous exponent or the trapping area, from the trajectories.

5.2. Nanoscale localization

Fluorescent probes frequently used in single-particle experiments applied to live biological samples include organic dyes, fluorescent proteins, quantum dots and other nanoparticles, etc., with distinct advantages (Cognet et al. 2014). Fluorescence microscopy is extremely sensitive because fluorescent emitters are excited at a specific wavelength, but emit at a different wavelength. Thus, the excitation light can be efficiently filtered away, allowing tiny amounts of fluorescence to become detectable on a dark background. Fluorescent probes can thus be nanometric. Usually, probes are targeted to specific organelles or proteins of interest (e.g. using antibodies, protein tags, chemical linking or protein fusion), to discriminate them against the rest of the sample. However, untargeted probes can also be observed while freely diffusing in the cellular or tissue environment, in which case they can yield information about the structure and rheology of that environment.

Fluorescent probes used to observe nanoscale processes are preferentially themselves nano-objects, in order not to bias the observation. Due to the limited resolution of optical microscopy, they can therefore be considered to be point emitters. When a single emitter is imaged (far from other emitters), the spread of its image is given by the point-spread function (PSF) of the microscope. When the sample contains a dense population of fluorescent markers, the size of this PSF then determines the resolution of a wide-field fluorescence microscope, known as the diffraction limit, $d = \lambda/(2\,\mathrm{NA})$ (where λ is the fluorescence wavelength and NA is the numerical aperture of the microscope objective).

In a single-particle tracking experiment, only a few particles are systematically imaged at a time so that it is possible to isolate each particle and determine their positions (the centers of their PSFs) with a resolution better than the diffraction limit. For example, the super-localized position of the PSF center can be obtained by fitting the experimentally measured image of a particle to a model PSF, or an approximation thereof, such as a 2D Gaussian:

$$I_{\mathrm{model}}(x, y; x_0, y_0, A, \sigma, I_b) = I_b + A \exp\left[-\frac{(x-x_0)^2 + (y-y_0)^2}{2\sigma^2}\right]. \qquad [5.1]$$

Here, (x, y) is the pixel coordinate, (x_0, y_0) is the center of the PSF, A is the amplitude of the PSF, σ is its standard deviation, and I_b is the background level (typically assumed to be locally constant). In practice, given an experimental image $I(x, y)$, the fit procedure consists of determining the values of x_0, y_0, A, σ and I_b that minimize the squared difference between the image and the model, $\sum_{x,y}(I(x, y) - I_{\text{model}}(x, y))^2$. This is a classical least-squares minimization problem, for which fast and robust implementations, such as the Levenberg–Marquardt method, are available (Press et al. 2007). The fitted (x_0, y_0) PSF center is then taken as the particle's position (see Figure 5.1). The image of a point source imaged via a high-NA objective is actually neither a Gaussian nor an Airy disc, but is well approximated by a Gaussian (Stallinga and Rieger 2010).

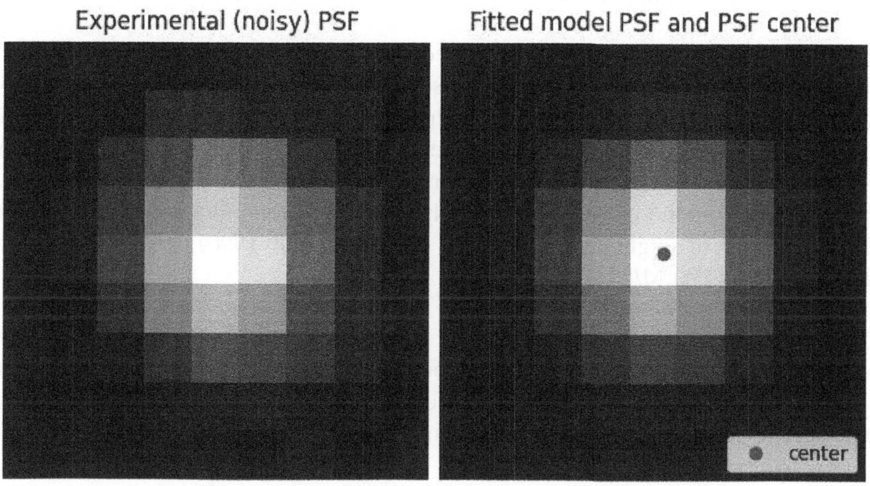

Figure 5.1. *Example of an experimental (noisy) PSF and the associated Gaussian fit and fitted PSF center*

In practice, a resolution $\sqrt{\langle(\Delta x)^2\rangle} \sim$ 10–20 nm is commonly attainable by such fits. Let s be the standard deviation of the PSF, N_p be the number of collected photons and a be the pixel size of the camera. There are three main contributions to the mean-square localization error (Mortensen et al. 2010):

$$\langle(\Delta x)^2\rangle \approx \underbrace{\frac{16}{9}\frac{\tilde{s}^2}{N_p}}_{(a)} + \underbrace{\frac{8\pi\tilde{s}^2 b^2}{a^2 N_p^2}}_{(b)} \quad \text{where} \quad \tilde{s}^2 = s^2 + \underbrace{\frac{a^2}{12}}_{(c)}, \qquad [5.2]$$

where:

– (a) is the shot (Poisson) noise, which, by the central limit theorem, scales as $1/\sqrt{N_p}$;

– (b) is the background noise (e.g. camera readout, or sample autofluorescence, modeled as a Gaussian of standard deviation b);

– (c) is the pixelation noise, equal to the spatial variance of a square pixel of size a.

A noteworthy point arising from this equation is that optimal localization is not achieved by minimizing the pixel size, but rather by making the pixel size match (approximately) the PSF size. Intuitively, this trade-off occurs because small pixel sizes decrease the pixelation error, but large pixel sizes average out the background noise. This effect can be related to the use of quadrant photodiodes (with only four pixels in total) to measure the position of laser beams, where the x position of the beam is obtained from the ratio between the left and right intensities, and likewise the y position from the ratio between the top and bottom intensities.

The localization methodology described here, initially applied for single-particle tracking, has been transposed to super-resolved localization microscopy techniques such as photoactivated localization microscopy (PALM) and stochastic optical reconstruction microscopy (STORM). The important difference, however, is that in single-particle tracking, photoactivation or photoswitching is typically not required, because the fluorophores are often naturally sparse. It is therefore not surprising that the first reports of single-particle tracking (Morrison et al. 1990) largely predate the development of PALM/STORM.

Because the basic super-localization algorithm is relatively simple, they are implemented in a very large number of publicly available software packages, with varying degrees of sophistication. We refer the reader to Sage et al. (2019) for an extensive review and comparison of these implementations.

5.3. Trajectory reconstruction

As mentioned previously, a single-particle tracking analysis pipeline can also reuse the high number of particle detection and localization routines recently developed for super-resolved localization microscopy. From these routines, we obtain a list of particle localizations $(\mathbf{r}_1, \mathbf{r}_2, \dots)$ on each frame $(t = 1, 2, \dots)$. The next step, however, is specific to single-particle tracking: these localizations must be *linked* together into trajectories. It is this process that we describe below.

The problem of linking localizations into trajectories is not unique to single-particle tracking. Even restricting ourselves to microscopy of biological samples, another instance of the problem occurs when studying cell motility, i.e. tracking the motion of cells. Some challenges are shared between the cell and particle tracking. For example, cells may divide in the middle of the experiment. Likewise, oligomeric protein complexes may undergo splitting. However, other difficulties are specific to single-particle tracking. In particular, single particles are typically indistinguishable from one another. Additionally, they may (again, depending on the system studied) frequently move in and out of the depth of field, particles may blink (i.e. disappear in some frames before reappearing later), and complexes may also undergo merging.

Naturally, the simplest case for linking occurs when only one particle is present in the entire field of view: in that case, there is a single localization on each frame (or none, if the particle can disappear), and all localizations can be directly connected to one another. Such a condition is normally easily achieved by working at a low enough fluorophore density, but has the important downside of greatly limiting the experiment throughput.

To handle the case of multiple particles present in the field of view, we will start by assuming that all particles are present over all frames, never appearing or disappearing in the middle of the experiment. We will also only consider the case of linking over two consecutive frames: in other words, we have a list of localizations $\mathbf{r}_1, \dots, \mathbf{r}_N$ on the first frame and another list of localizations $\mathbf{r}'_1, \dots, \mathbf{r}'_N$ on the second frame, and we seek to pair the localizations, or equivalently to order the second list, so that the particle at $\mathbf{r}_1$ is moved to $\mathbf{r}'_1$, the particle at $\mathbf{r}_2$ is moved to $\mathbf{r}'_2$, etc. Linking over the entire movie will then be achieved by linking the first frame to the second one, the second to the third, etc.

Given such a pairing, the displacement vectors of the particles are $\mathbf{r}'_1 - \mathbf{r}_1, \dots, \mathbf{r}'_N - \mathbf{r}_N$. We will show in section 5.4 that in many biologically relevant situations, the probability density $p(\Delta\mathbf{x})$ that a particle moves by a vector $\Delta\mathbf{x}$ during the inter-frame time Δt is given by

$$p(\Delta\mathbf{x}) = \frac{1}{(4\pi Dt)^{d/2}} \exp\left(-\frac{|\Delta\mathbf{x}|^2}{4D\Delta t}\right) \qquad [5.3]$$

where D is the *diffusion coefficient* of the particle and d is the dimensionality of the trajectory (commonly, $d = 2$, or $d = 3$ for 3D localization, not reviewed above). The probability density associated with the given displacement vectors is the product of individual probabilities, i.e.

$$p(\{\mathbf{r}_1, \dots, \mathbf{r}_N, \mathbf{r}'_1, \dots, \mathbf{r}'_N\}) = \frac{1}{(4\pi D\Delta t)^{dN/2}} \exp\left(-\frac{|\mathbf{r}'_1 - \mathbf{r}_1|^2 + \dots + |\mathbf{r}'_N - \mathbf{r}_N|^2}{4D\Delta t}\right); \qquad [5.4]$$

maximizing this probability amounts to minimizing the sum of the particles' squared displacements (Crocker and Grier 1996).

A priori, in order to determine the correct pairing of localizations, we need to consider all $N!$ possible localization pairings. In practice, such an exhaustive search would quickly become overly expensive computationally. Therefore, a cutoff displacement d_{max} is normally used: we assume that no particle can move by more than that distance over the inter-frame time Δt. This cutoff displacement can, for example, be determined based on the decay of the displacement distribution (equation [5.3]). Unless the particles are very densely distributed, this cutoff only leaves a few pairings to consider, over which equation [5.4] can be maximized.

Particles that appear or disappear in the middle of the experiment manifest themselves as localizations that cannot be paired with a localization in the preceding or following frame, respectively. If we take into account the displacement cutoff, they correspond, for example, to a single localization that does not have a neighboring-frame localization at a distance less than d_{max}, or to two close localizations on the same frame which altogether only have a single neighboring-frame localization at a distance less than d_{max}, etc. In that case, we can maximize equation [5.4] over the pairings that include all localizations on the frame with fewer localizations, and interpret the remaining localizations (on the frame with more localizations) as corresponding to appearing or disappearing trajectories.

As for localization, various software packages implementing trajectory reconstruction are publicly available. We refer the reader, for example, to Tinevez et al. (2017) (TrackMate) for an example.

5.4. Nanoscale dynamics

Having localized, with nanometer precision, single particles moving in biological samples, and reconstructed their trajectories, we would now like to infer biophysical parameters from these trajectories.

5.4.1. *Basic properties of Brownian motion*

The position $\mathbf{x}$ of a spherical particle of mass m and radius R, in a fluid of viscosity η and driven by thermal fluctuations is described by the Langevin equation,

$$m\ddot{\mathbf{x}} = -6\pi\eta R\dot{\mathbf{x}} + \varepsilon, \qquad [5.5]$$

where dots indicate derivatives with respect to time. This equation takes into account the two forces exerted on the particle. The first term is the viscous (Stokes) drag $-6\pi\eta R\dot{\mathbf{x}}$. Even for nonspherical particles, an effective R (the *hydrodynamic radius*) can be used so that this equation is satisfied. The second term is the thermal force ε, which is a random noise with no correlation in time or in any spatial coordinates, i.e.

$$\langle \varepsilon_i(t)\varepsilon_j(t')\rangle \propto \delta_{i,j}\delta(t-t'). \qquad [5.6]$$

Order of magnitude calculations show that nanoscale motion in biological samples always occurs in the so-called low-Reynolds regime, where the inertial term $m\ddot{\mathbf{x}}$ is negligible, and equation [5.5] reduces to $\dot{\mathbf{x}} = \varepsilon/6\pi\eta R$, again with $\dot{\mathbf{x}}$ being entirely uncorrelated.

Because the Langevin equation involves a random noise component, we cannot obtain a deterministic solution to it. However, we can compute the expected value or the probability distribution of some functions of the particle position.

The mean-squared displacement (MSD) of the particle is one such function that can be computed. After N "ticks" of a short time interval Δt (i.e. a *time lag* of $N\Delta t$), the MSD satisfies

$$\begin{aligned}
\mathrm{MSD}(t) &= \left\langle \left[\mathbf{x}(N\Delta t) - \mathbf{x}(0)\right]^2 \right\rangle \\
&= \left\langle \left[\left(\mathbf{x}(\Delta t) - \mathbf{x}(0)\right) + \cdots + \left(\mathbf{x}(N\Delta t) - \mathbf{x}((N-1)\Delta t)\right)\right]^2 \right\rangle \\
&\overset{\Delta t \to 0}{=} \left\langle \left[\left(\dot{\mathbf{x}}(0) + \cdots + \dot{\mathbf{x}}((N-1)\Delta t)\right)\Delta t\right]^2 \right\rangle \\
&= N\left\langle \dot{\mathbf{x}}^2 \right\rangle \Delta t^2
\end{aligned} \qquad [5.7]$$

where the last equality is established by expanding the square and noting that all of the cross-terms have zero mean. In particular, this means that the MSD grows linearly with N, i.e. with time. The rate of this growth defines the *diffusion coefficient D*, namely

$$\mathrm{MSD}(t) = 2dDt \qquad [5.8]$$

where d is the dimensionality of the system (two for 2D motion, three for 3D motion).

The value of the diffusion coefficient is then derived by combining equations [5.5] and [5.6] and the equipartition theorem ($\langle \frac{1}{2} m\mathbf{v}^2 \rangle = \frac{d}{2} k_B T$, where k_B is the Boltzmann constant, and T is the temperature): we obtain

$$D = \frac{k_B T}{6\pi\eta R}. \qquad [5.9]$$

Because the temperature is typically well controlled, variations in the diffusion coefficient D imply variation in either the viscosity (as felt by the particle) or in the particle's hydrodynamic radius (e.g. molecular interactions, protein dimerization, etc.).

We can also estimate the probability of observing any vector displacement $\Delta\mathbf{x}$ after a time $N\Delta t$: this is

$$\begin{aligned}
&p(\mathbf{x}(N\Delta t) - \mathbf{x}(0) = \Delta\mathbf{x}) = \\
&\qquad p\left(\left(\dot{\mathbf{x}}(0) + \cdots + \dot{\mathbf{x}}((N-1)\Delta t)\right)\Delta t = \Delta\mathbf{x}\right)
\end{aligned} \qquad [5.10]$$

which is the probability distribution of a sum of N independent and identically distributed variables. The central limit theorem states that this distribution converges towards a normal distribution,

$$p(\Delta \mathbf{x}) = \frac{1}{(4\pi Dt)^{d/2}} \exp\left(-\frac{|\Delta \mathbf{x}|^2}{4D\Delta t}\right) \qquad [5.11]$$

(where the scaling factor can be determined from equation [5.8]). In particular, in two dimensions, this result implies that the probability of observing a displacement of length ℓ (independently of its orientation) is

$$p(\ell) = \frac{\ell}{2Dt} \exp\left(-\frac{\ell^2}{4Dt}\right) \qquad [5.12]$$

(a 2D Maxwell–Boltzmann distribution).

5.4.2. *Quantification of experimental trajectories*

Given an experimental trajectory observed at discrete times $\mathbf{x}_1, \ldots, \mathbf{x}_N$, equations [5.11] and [5.12] provide a (necessary, but not sufficient) condition to check whether the trajectory arises from the Brownian motion: the distribution of frame-to-frame displacements $\mathbf{x}_2 - \mathbf{x}_1, \ldots, \mathbf{x}_N - \mathbf{x}_{N-1}$ should be normally distributed. The distribution of displacement lengths $|\mathbf{x}_2 - \mathbf{x}_1|, \ldots, |\mathbf{x}_N - \mathbf{x}_{N-1}|$ should follow a Maxwell–Boltzmann distribution. The latter requirement is less stringent, but easier to check, simply by histogramming the displacement lengths.

Having verified this condition, equation [5.8] then provides a simple method to estimate the diffusion coefficient of the particle: the MSD can be computed at each time lag by averaging the squared displacements at that lag, i.e.

$$\begin{aligned}
\mathrm{MSD}(t) &= \frac{1}{N-1}\left[|\mathbf{x}_2 - \mathbf{x}_1|^2 + \cdots + |\mathbf{x}_N - \mathbf{x}_{N-1}|^2\right], \\
\mathrm{MSD}(2t) &= \frac{1}{N-2}\left[|\mathbf{x}_3 - \mathbf{x}_1|^2 + \cdots + |\mathbf{x}_N - \mathbf{x}_{N-2}|^2\right], \\
&\vdots \\
\mathrm{MSD}((N-1)t) &= \frac{1}{1}\left[|\mathbf{x}_N - \mathbf{x}_1|^2\right]
\end{aligned} \qquad [5.13]$$

where t is the frame duration. We can then plot the MSD versus the time lag and check whether it indeed grows linearly (see section 5.4.4 for an explanation of the case where it will not be treated). If so, a linear fit yields the slope of the growth, equal to $2dD$.

However, there are two practical points that need to be taken into account here (Michalet and Berglund 2012). On the one hand, the presence of localization error and finite exposure times modifies the MSD equation. On the other hand, the fact that measured trajectories have finite length implies that MSDs at long time lags are poorly estimated, and need to be removed from the linear fit.

5.4.2.1. *Effect of localization error on the MSD*

Let us first consider the effect of localization error on the MSD. Let $\mathbf{x}(t)$ be the true particle localization at any time t. In reality, the particle is not observed continuously, but over discrete frames. To model this effect, we will assume that the observed particle position is the mean of its true position over the exposure time t_E. In addition, this position is corrupted by a localization error $\Delta\mathbf{x}$. Overall, the observed particle position $\mathbf{x}_{\text{obs}}$ is therefore

$$\mathbf{x}_{\text{obs}} = \langle \mathbf{x}(t) \rangle_{\text{frame}} + \Delta\mathbf{x} \qquad [5.14]$$

where $\Delta\mathbf{x}$ is described in equation [5.2]. In this model (Michalet and Berglund 2012), we can show that the MSD of the observed particle position is no longer $\text{MSD}(t) = 2dDt$, but

$$\text{MSD}_{\text{obs}}(t) = 2dD\left(t - \frac{4}{3}t_E\right) + 2d\left\langle(\Delta x)^2\right\rangle. \qquad [5.15]$$

In other words, finite exposure time shifts the MSD curve rightwards by $\frac{4}{3}t_E$, whereas localization uncertainty shifts it upwards by $2d\left\langle(\Delta x)^2\right\rangle$. The slope, however, is unmodified. Therefore, the linear fit proposed above simply needs to be modified to include a nonzero intercept and the procedure remains valid. From the fitted intercept (equal to $2d(\left\langle(\Delta x)^2\right\rangle - \frac{4}{3}Dt_E)$), we can extract the mean localization error $\left\langle(\Delta x)^2\right\rangle$ (as the exposure time t_E is known), which can be compared with the estimate from equation [5.2] for verification.

5.4.2.2. *Effect of finite trajectory length on the MSD*

The second issue that needs attention is the finite length N of the trajectory. Equation [5.13] evaluates the MSD at lag n by averaging $N - n$ squared displacements. Therefore, the larger the lag, the fewer values are averaged. Additionally, the noise that corrupts each squared displacement term becomes correlated, which makes the usual $\sqrt{N-n}$ weights invalid. Careful analyses show that the diffusion coefficient can be best estimated using an unweighted linear fit restricted to a small number of time lags, which depends on the relative values of the static localization error $\left\langle(\Delta x)^2\right\rangle$ and the mean-square displacement after one frame, $2dD\Delta t$, but may be as low as two lags for fast-moving particles. Complete procedures for determining the optimal number of time lags in a data-dependent manner have been described in the literature (Michalet and Berglund 2012). It has also been shown that the relative uncertainty in the diffusion coefficient obtained by this procedure is, at best (if the static localization error is small) on the order of $\sqrt{6/dN}$ (for example, for 2D trajectories, 50 points allow estimating D with an $\sim 25\,\%$ precision and 300 points yield a $10\,\%$ precision (Michalet and Berglund 2012)).

5.4.3. *Time-variable diffusion*

So far, we have described the motion of a particle undergoing Brownian motion with a constant diffusion coefficient. However, when tracking, for example, proteins in inhomogeneous environments, such as cells and tissues, changes in diffusion coefficient commonly occur over time and space. Such changes can sometimes be linked to specific biological processes, for example, diffusion can be slowed down by temporary protein–protein binding that increase the effective size of the tracked particle, or other protein–protein interactions that temporarily trap the tracked particle. In that case, the change in diffusion coefficient is not random. Often, the particle simply switches between a small number of *hidden* (i.e. not directly observable) states (e.g. states (1), (2) or (3), each associated with its diffusion coefficients D_1, D_2, D_3). Each state can correspond, for example, to different protein binding states or well-defined local environments. The presence of such states with different diffusion coefficients can be diagnosed, for example, if the distribution of displacement lengths (equation [5.12]) is not well fitted by a Maxwell–Boltzmann distribution, but by a mixture of two or more Maxwell–Boltzmann distributions.

Another condition (the *Markov condition*) is often applicable. It states that for each frame, the particle may stay in the state it was previously, or transition to a new state, with probabilities that only depend on the original and new state ($p_{1\to 2}$, $p_{2\to 1}$, $p_{2\to 3}$, $p_{3\to 2}$, $p_{3\to 1}$, $p_{1\to 3}$), but not, for example, on how much time the particle has spent in the state. These probabilities depend on the frame rate, but can be equivalently expressed as *transition rates* ($k_{i\to j} \approx p_{i\to j}/\Delta t$), which are independent of the frame rate. In that case, the particle's state switching is described as a so-called *hidden Markov model* (HMM). The goal of the analysis is then to infer the number of underlying states, their associated diffusion coefficients and the transition probabilities (in the hidden Markov models study, we commonly also aim to infer the hidden state at each time point of the trajectory, but this task is typically not required in the context of single-particle tracking). The actual inference procedure is beyond the scope of this review, but have been described in the literature, and again, implementations are publicly available (e.g. Persson et al. 2013).

5.4.3.1. *Example: modulation of the dynamics of Hfq under rifampicin treatment*

As a sample application, we review here the analysis of the effect of rifampicin (an antibiotic which inhibits DNA transcription) on the RNA binding protein Hfq, performed in Persson et al. (2013). Briefly, trajectories of fluorescently labeled Hfq diffusing in *E. coli* were recorded in the absence and the presence of rifampicin, and analyzed using an HMM-based algorithm (see Figure 5.2). It was found that Hfq natively switched between three different diffusive states (slow: $D_1 = 0.24\,\mu\mathrm{m}^2/\mathrm{s}$; medium: $D_2 = 0.75\,\mu\mathrm{m}^2/\mathrm{s}$; fast: $D_3 = 2.6\,\mu\mathrm{m}^2/\mathrm{s}$). Rifampicin treatment removed the slowest state and decreased the rate of transition from the fast state to the medium state as well as the population of the medium state (and slightly altered the corresponding diffusion coefficients). These results were interpreted as follows: the fast state corresponds to free Hfq, whereas the medium and slow states correspond to RNA-bound Hfq; the slow state, which disappears under rifampicin treatment, corresponds to Hfq bound to RNA being transcribed (rifampicin inhibits transcription). The decrease in the fast-to-medium transition rate is due to the decrease in RNA to which Hfq can bind. On the other hand, the dissociation rate of Hfq from RNA should not be affected by rifampicin, explaining the lack of change in the reverse, medium-to-fast transition rate.

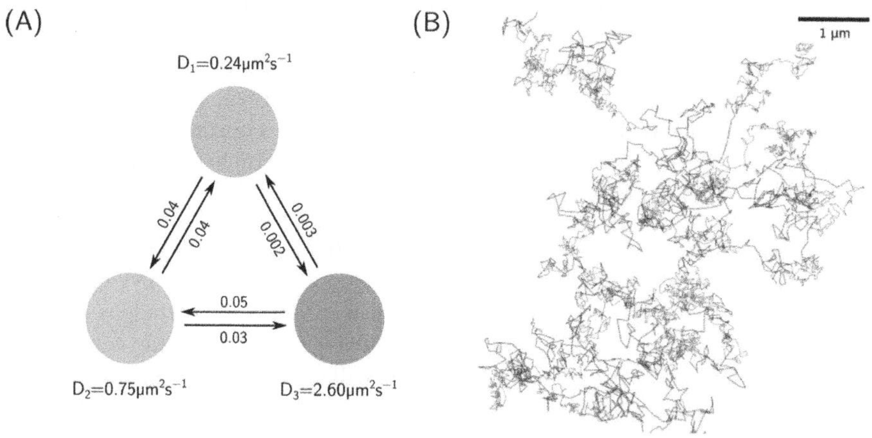

Figure 5.2. *Illustration of the hidden Markov model description of Hfq diffusion in* E. coli, *proposed by Persson et al. (2013). (A) Three-state model of Hfq diffusion. Each state is associated with a diffusion coefficient, and transition probabilities are given for each pair of states (here, computed for a frame rate of 300 Hz). (B) Simulated trajectory associated with this diffusive behavior. Individual segments have been colored according to the underlying diffusive state: the simulated Hfq visibly switches between slow (blue), medium (green) and fast (red) diffusion*

5.4.4. *Beyond Brownian motion: anomalous diffusion*

A situation where the Brownian description of diffusion breaks down to a larger extent occurs when the individual frame-to-frame displacements (equation [5.10]) are not independent of one another, i.e. there is some "memory" effect. If the memory only lasts for a few frames, then we can group these frames together (i.e. pretend that the movie was recorded at a lower frame rate), and still describe the resulting, coarsened trajectory using the Brownian motion. However, there are also cases where the memory lasts much longer (technically, this usually corresponds to power-law autocorrelation decays, $\langle \Delta x_t \Delta x_{t+u} \rangle_t \sim Ct^\nu$). In that case, the particle's behavior becomes very different. In particular, the mean-square displacement no longer grows linearly with time (equation [5.8]), but as a power of time:

$$\mathrm{MSD}(t) \propto t^\alpha \quad ; \quad \alpha \neq 1. \qquad [5.16]$$

The exponent α is known as the *anomalous exponent*. The theory of anomalous diffusion is relatively complex, but a simple semi-quantitative description can be achieved by simply computing the MSD and observing its deviations from linearity.

Two different cases occur, depending on the sign of the displacements' autocorrelation (i.e. the memory). On the one hand, if the autocorrelation is positive, once the particle starts moving in a direction, it tends to keep moving in that direction. This behavior can describe, for example, particles in a flow, or actively carried by directional molecular motors. In that case, the MSD grows super-linearly, i.e. $\alpha > 1$. In particular, in the case of pure directional motion at a constant velocity $\mathbf{v}$, we have $\mathbf{x}(t) = \mathbf{v}t$ and thus $\mathrm{MSD}(t) = v^2t^2$ (i.e. the MSD grows quadratically ($\alpha = 2$)). Intermediate cases ($1 < \alpha < 2$) occur, for example, when particle motion is a mixture of directed motion and Brownian diffusion.

Conversely, the autocorrelation can be negative: after moving in a certain direction, the particle is more likely to take future steps in the opposite direction. In general, this is actually an effective property of the environment in which the particle is moving: if this environment contains many impassable "walls" (which commonly occur in the crowded cell or tissue environments), the particle will tend to "bump" into these walls and therefore move backwards. In that case, we may observe *anomalous subdiffusion*, i.e. a sublinear mean-square displacement ($\alpha < 1$).

A more extreme case and qualitatively different case of negative autocorrelation is corralled diffusion, which occurs when the particle is completely trapped by these walls in a finite region, from which it can never escape. In this case, the mean-square displacement cannot even diverge, but is bounded by a quantity on the order of the square of the trapping region diameter L, typically converging to it exponentially (Kusumi et al. 1993):

$$\mathrm{MSD}(t) \approx \frac{L^2}{3} \times (1 - \exp(-12Dt/L^2)) \qquad [5.17]$$

Fitting the empirical MSD to this formula therefore yields both the diffusion coefficient D and the trapping region diameter L.

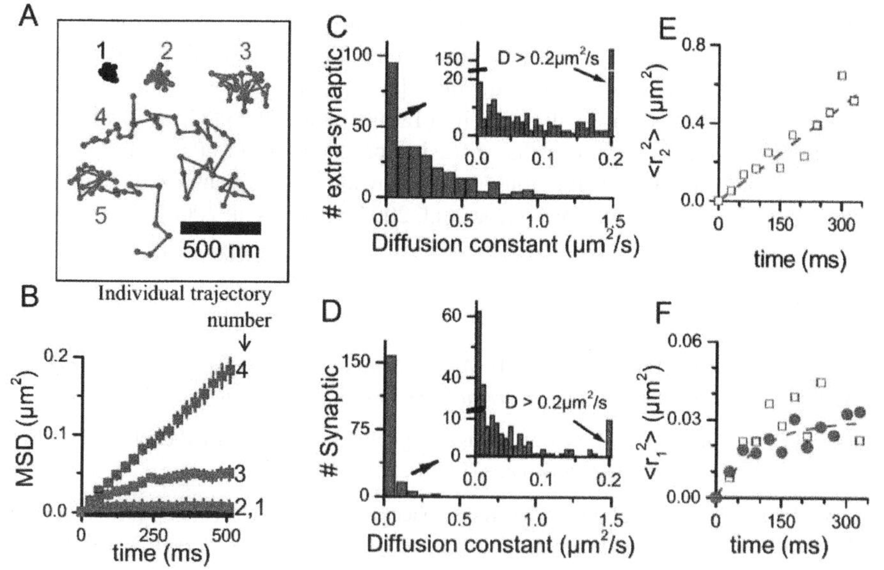

Figure 5.3. *(A) Sample trajectories of a fixed fluorophore (black), and of synaptic (green) and extra-synaptic (red) labeled neuroreceptors. (B) Mean-square displacements of the trajectories shown in (A), demonstrating constrained synaptic diffusion and free extra-synaptic diffusion. (C) and (D) Distributions of diffusion coefficients for extra-synaptic and synaptic receptors: synaptic receptors diffuse much more slowly than extra-synaptic ones. (E) and (F) MSDs were split into a constrained component ($\langle r_1^2 \rangle$) and a free component ($\langle r_2^2 \rangle$). Synaptic receptors (green) exclusively displayed confined diffusion, whereas extra-synaptic ones (red) displayed a mixture of unbounded and of confined diffusion. Modified with permission from Tardin et al. (2003)*

5.4.4.1. *Example: dynamics of membrane receptor movements in the synapses of live neurons*

As an example application, we discuss here the dynamics of neurotransmitter receptors, which play a central role in synaptic transmission and in memory processes. Single-particle tracking has been instrumental in elucidating the diffusion of such receptors in and out of synapses. Without single molecule imaging, wide-field fluorescence observations of receptor dynamics are hampered by the small size of the synapses ($\sim$ 300 nm), close to the diffraction limit. Therefore, it was previously thought that receptors were immobile at the synapse, and that they exchanged via trafficking, i.e. exocytosis of naive receptors and endocytosis of desensitized ones. Single-particle tracking of receptors has shown that, in fact, they could diffuse laterally in and out of the synapse, and could also be mobile within the synapse. More precisely, trajectory analysis revealed different diffusive regimes depending on the receptor location: whereas extra-synaptic diffusion (i.e. outside of synapses) could be unbounded or confined, receptor diffusion in post-synaptic membranes (i.e. in synapses) was exclusively confined (see Figure 5.3). The dimension of confinement, as extracted using equation [5.17], matched structural (electron microscopy) data of synapse dimensions (Tardin et al. 2003; Groc et al. 2004). Furthermore, it was later shown that diffusion processes inside the synapse play a key role in fast synaptic transmission processes by allowing desensitized receptors to be quickly replaced by naive ones (Groc et al. 2004; Heine et al. 2008).

5.5. References

Cognet, L., Leduc, C., Lounis, B. (2014). Advances in live-cell single-particle tracking and dynamic super-resolution imaging. *Current Opinion in Chemical Biology*, 20, 78–85.

Crocker, J.C. and Grier, D.G. (1996). Methods of digital video microscopy for colloidal studies. *Journal of Colloid and Interface Science*, 179(1), 298–310.

Groc, L., Heine, M., Cognet, L., Brickley, K., Stephenson, F.A., Lounis, B., Choquet, D. (2004). Differential activity-dependent regulation of the lateral mobilities of AMPA and NMDA receptors. *Nature Neuroscience*, 7(7), 695–696.

Heine, M., Groc, L., Frischknecht, R., Béïque, J.-C., Lounis, B., Rumbaugh, G., Huganir, R.L., Cognet, L., Choquet, D. (2008). Surface mobility of postsynaptic ampars tunes synaptic transmission. *Science*, 320(5873), 201–205.

Kusumi, A., Sako, Y., Yamamoto, M. (1993). Confined lateral diffusion of membrane receptors as studied by single particle tracking (nanovid microscopy). Effects of calcium-induced differentiation in cultured epithelial cells. *Biophysical Journal*, 65(5), 2021–2040.

Michalet, X. and Berglund, A.J. (2012). Optimal diffusion coefficient estimation in single-particle tracking. *Physical Review E*, 85(6), 061916.

Morrison, I.E.G., Anderson, C.M., Georgiou, G.N., Cherry, R.J. (1990). Measuring diffusion coefficients of labelled particles on cell surfaces by digital fluorescence microscopy. *Biochemical Society Transactions*, 18(5), 938–938.

Mortensen, K.I., Churchman, L.S., Spudich, J.A., Flyvbjerg, H. (2010). Optimized localization analysis for single-molecule tracking and super-resolution microscopy. *Nature Methods*, 7(5), 377–381.

Persson, F., Lindén, M., Unoson, C., Elf, J. (2013). Extracting intracellular diffusive states and transition rates from single-molecule tracking data. *Nature Methods*, 10(3), 265–269.

Press, W.H., Teukolsky, S.A., Vetterling, W.T., Flannery, B.P. (2007). *Numerical Recipes 3rd Edition: The Art of Scientific Computing*. Cambridge University Press, Cambridge.

Sage, D., Pham, T.-A., Babcock, H., Lukes, T., Pengo, T., Chao, J., Velmurugan, R., Herbert, A., Agrawal, A., Colabrese, S. et al. (2019). Super-resolution fight club: Assessment of 2D and 3D single-molecule localization microscopy software. *Nature Methods*, 16(5), 387–395.

Stallinga, S. and Rieger, B. (2010). Accuracy of the Gaussian point spread function model in 2D localization microscopy. *Optics Express*, 18(24), 24461–24476.

Tardin, C., Cognet, L., Bats, C., Lounis, B., Choquet, D. (2003). Direct imaging of lateral movements of AMPA receptors inside synapses. *The EMBO Journal*, 22(18), 4656–4665.

Tinevez, J.-Y., Perry, N., Schindelin, J., Hoopes, G.M., Reynolds, G.D., Laplantine, E., Bednarek, S.Y., Shorte, S.L., Eliceiri, K.W. (2017). Trackmate: An open and extensible platform for single-particle tracking. *Methods*, 115, 80–90.

6

In Depth Microscopy

Tom DELAIRE[1,2,3] **and Rémi GALLAND**[1]

[1] *Interdisciplinary Institute for Neuroscience, UMR 5297, CNRS, University of Bordeaux, France*

[2] *Department of Experimental Medical Science, Lund University, Sweden*

[3] *Lund Bioimaging Center, Lund University, Sweden*

6.1. Introduction

Research in biology builds on the observation of many diverse biological samples and models ranging from single isolated cells ($\approx 10^3$ μm^3) to whole organisms and animals (up to mm^3) (see Figure 6.1(A)). In that quest, fluorescence light microscopy offers cell biologists the unique capability of monitoring, with a very high specificity, virtually any proteins or bio-molecules in conditions similar to their native environments (Giepmans et al. (2006) and Chapter 3). Nowadays, there is a major trend towards imaging models that are increasingly physiologically relevant, which implies observing (1) bigger, (2) more complex, and ideally (3) living samples.

Fluorescence microscopy's ability to image in depth within 3D samples is governed by two main factors: its spatial resolution and the optical properties of the sample itself. The spatial resolution is inherently driven by the phenomenon of light diffraction, for which a point source located in the object plane results from the formation of a diffraction pattern onto the camera called the point spread function (PSF). Its volume depends on the objective numerical aperture (NA), light wavelength (λ) and the imaging media refraction index (n), and rapidly diverges in the axial direction (see Figure 6.1(B)). It follows that any emitter outside the microscope objective's focal plane will produce a fuzzy halo on the image, leading to a possibly disastrous loss of contrast for the in-focus structures. It defines the depth of field (DoF)

Photonic Imaging for Biology,
coordinated by Jean-Baptiste SIBARITA.

over which a microscope is able to produce a clear image of the sample (see Figure 6.1(C)).

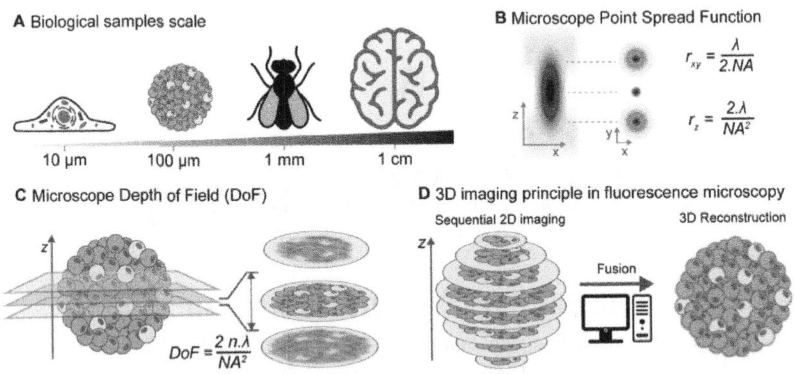

Figure 6.1. *A. Typical length scale of biological sample that fluorescence microscopy aims to image for biological studies. B. Point spread function of a fluorescence microscope which defines its radial (r_{xy}) and axial (r_z) resolutions, as well as the depth of field over which a microscope will produce a net image of the sample (C). D. For 3D imaging, a set of 2D optically sectioned planes are acquired and then fused together to produce a 3D representation of the imaged sample. For a color version of this figure, see www.iste.co.uk/sibarita/photonic.zip*

Imaging 3D objects therefore relies on the capacity to effectively reject any out of focus light and only retain the in-focus information. This ability, usually referred to as *optical sectioning*, forms the foundations of 3D microscopy. A sample can then be acquired sequentially one plane at a time along the objective optical axis in order to get a complete and clear 3D representation of it (see Figure 6.1(D)). However, this process requires successive illuminations and signal collections, limiting the imaging speed and possibly inducing deleterious photo-toxic effects to the sample.

Beyond resolution, the achievable penetration depth of an imaging method inside a biological sample is governed by its optical properties. This depth is usually approximated by the mean free path a photon can travel before being absorbed or scattered by a molecule (≈ 100 µm up to mm in the visible and the NIR range of the spectrum, respectively (Yaroslavsky et al. 2002)), above which any structure becomes hardly discernible. Lastly, sample complexity, represented by the inhomogeneities of its refractive indices, induces depth-dependent optical aberrations. These aberrations broaden the microscope's PSF and impact its spatial resolution. These characteristics of the sample can lead to a complete blurring of the observed structure, representing the ultimate challenge for in-depth microscopy.

In this chapter, we will provide an overview of the different optical sectioning methods that enable performing 3D imaging of possibly complex samples. We will discuss their benefits and limitations regarding their penetration depth, their imaging speed, their photo-toxic effect or their complexity. Focusing on fluorescence imaging, we will first dive into approaches that make it possible to provide an in-depth optical sectioning through spatial filtering (see section 6.2) or excitation light confinement (sections 6.3 and 6.4), followed by alternative solutions based on image post-processing (section 6.5). Finally, we will discuss actual challenges that are currently being adressed to continue the quest of imaging always more complex and physiologically relevant 3D living samples (section 6.6).

6.2. Confocal microscopy

6.2.1. *Principle*

Confocal laser scanning microscopy (CLSM) is the most widely used 3D imaging technique due to its impressive optical sectioning capacity and ease of use. The concept of confocal detection was first introduced in the early 1950s by Marvin Minsky, but its real success started in the 1980s with the advent of reliable and accessible laser sources, sensitive detectors and the development of commercial solutions (Wilson and Sheppard 1984).

CLSM 3D sectioning ability relies on focusing a laser beam through the objective followed by the spatial filtering of the excited out-of-focus signal through a pinhole inserted at the image plane in front of the detector. Consequently, only the light tightly focused at the pinhole plane, which corresponds to the light originating from the objective's focal plane, will contribute to the image formation (see Figure 6.2(A)). In comparison, in standard widefield microscopy, both in-focus and out-of-focus lights are collected by the camera leading to poorer contrast when 3D objects are considered. However, this spatial filtering solution is only possible when a single point is excited and detected, at a time. To build an image, a focused excitation spot of light has to be scanned throughout the sample. This is nowadays achieved thanks to a scanning unit generally composed of a pair of galvanometric mirrors, which moves the position of a focused laser spot across the sample. The signal is then de-scanned passing back through the scanning unit and focused onto a stationary pinhole for spatial filtering before being detected by a single point detector (photodiode, photomultiplier tube, avalanche photodiode) (see Figure 6.2(A)).

In CLSM, the product of the excitation and the detection PSFs defines the confocal volume which actually contributes to the image formation, and whose size determines the spatial resolution that can be achieved. In theory, if we consider an infinitesimally small pinhole, the lateral and axial resolution of a confocal microscope can be improved up to a factor of 2 as compared to widefield microscopy. In practice, the induced signal loss limits its use and a trade-off has to be made in between the pinhole size, affecting the resolution and sectioning gain, and the number of collected photons. In fluorescence microscopy, where samples have a limited brightness, a pinhole having the size of the first annulus of the Airy pattern corresponding to the objective's NA (i.e. 1 Airy Unit) is commonly used. It is also important to consider the pixel size, namely the spacing in between two signal measurements, which is defined by the scanning speed and detection frequency. Bigger pixels will degrade the overall resolution, whereas too small pixels will lead to oversampling and excessive illumination dose onto the sample.

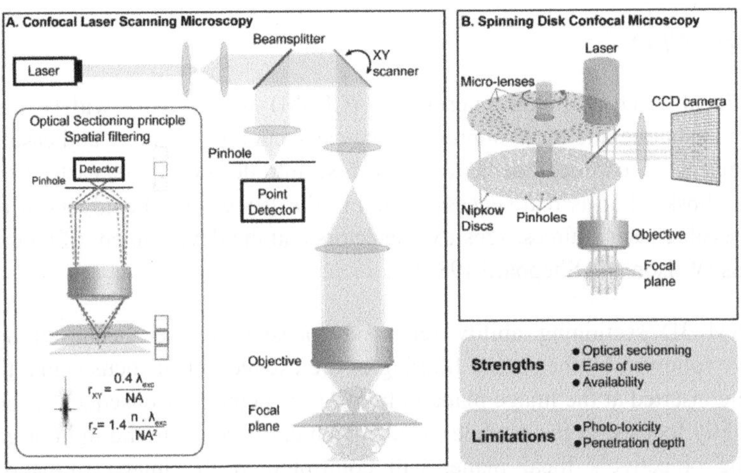

Figure 6.2. *A. Standard CLSM optical setup. The inset illustrates the spatial filtering principle using a pinhole which provides 3D optical sectioning and can potentially lead to improved spatial resolution. B. SDCM principle: two Nipkow discs enable the creation of multiple excitation foci on the sample and spatially filter the collected signal before being projected onto a 2D camera. A confocal image is built up when the disks are spun during an integration time of the camera*

Even though CLSM enables improving the spatial resolution as compared to conventional approaches, its strength mostly lies in its optical sectioning ability, when an out-of-focus signal has to be considered. Indeed, signal contribution from the out-of-focus plane drastically decreases with the axial distance in CLSM, as compared to widefield detection where it remains constant. This leads to very sharp and contrasted images of 2D sections even inside highly fluorescent 3D samples

(Conchello and Lichtman 2005; Wilson 2011). Therefore, confocal microscopy outperforms most of the 3D imaging techniques in terms of image contrast when it comes to image sample sizes in the range of 5–100 μm. For bigger samples, absorption, scattering, and optical aberrations will affect the achievable signal-to-noise ratio and thus ultimately limit the penetration depth.

6.2.2. *Spinning disk confocal microscopy*

A key limitation of CLSM is its imaging speed as a single excitation spot has to be scanned throughout all the sample to build a 2D image. Hence, CLSM generally requires seconds to acquire a full-frame image (512 x 512 and above), making it unsuitable to resolve many dynamics within cells and restrict the achievable imaging throughput. Important efforts have been made to increase this scanning speed using resonant galvanometric mirrors, faster acousto-optic deflectors or even confocal line detection approaches. Nevertheless, increasing the excitation laser scanning speed ultimately reaches another limitation, that is the time spent per pixel, referred as the dwell time. Indeed, faster scanning speed decreases the dwell time and therefore the amount of fluorescence signal that can be collected per pixel. This is due to the fluorescence lifetime, which corresponds to the average time it takes for a fluorophore after absorption to emit a photon (in the order of nanoseconds), which intrinsically limits the number of photons a fluorophore can emit in a given time. This eventually limits the achievable image to noise ratio for faster acquisition speed (Conchello and Lichtman 2005).

Spinning disk confocal microscopy (SDCM) provides an alternative method to increase CLSM speed, up to video rate, while largely preserving its optical sectioning capacity. As compared to CLSM, SDCM parallelizes the excitation foci, greatly decreasing the time required to acquire a single image without compromising the pixel dwell time. It builds on the so-called Nipkow disk, proposed in 1885 by Paul Gottlied Nipkow for mechanical image decomposition, and which is composed of equally spaced and identical circular holes arranged along spirals of constant pitch. Positioned in the laser excitation path, it provides hundreds of simultaneous excitation foci that scan across the sample, building up an image when the disk is spun. The signal is then collected and filtered out through the same disk to provide confocality, namely optical sectioning, and detected onto a 2D camera (see Figure 6.2(B)). Usually, a second disk made of an array of micro-lenses is added on top of the regular Nipkow disk to focus the laser light through the pinholes more efficiently. Combined with a high speed and sensitive camera for signal detection, SDCM makes it possible to build up images at a rate up to 100 s of frames per second, making it suitable to monitor fast dynamics within living samples. However, this increase of acquisition speed comes with a slight decrease in spatial resolution (equivalent to widefield microscopy) and sectioning capacity due to signal crosstalk in between the many pinholes composing the Nipkow disk. This effect increases in relation with the sample thickness.

6.2.3. *Limitations*

CLSM, and its cousin SDCM, are undoubtedly the most widely used 3D imaging systems, and many commercial and homemade systems are available. Yet, confocal microscopy still suffers from several limitations when it comes (1) to imaging living and light-sensitive samples, and (2) to imaging more than the first tens to hundreds of micrometers into a sample. It arises first from the non-optimal use of the excitation light, since absorption occurs all along the laser path, but only signals originating from the focal plane are actually used. Consequently, confocal imaging leads to important photo-induced damaging effects such as fast bleaching of the fluorophores both in and out of the imaged plane. Additionally, it leads to the creation of deleterious reactive oxygen species (ROS) from the excitation of exogenous and endogenous chromophores all along the laser path, greatly limiting its ability to perform live imaging. Secondly, only ballistic photons (i.e. photons that have not been scattered or absorbed on their path), both in excitation and in detection, effectively contribute to the signal, while scattered photons either decrease the excitation intensity or are filtered out in detection. The necessary increase in excitation intensity, to compensate for this scattering, further exacerbates photobleaching and photo-toxic effects. CLSM imaging penetration depth is thus rapidly limited in non-fully transparent samples. Nevertheless, due to its high sectioning capacity, ease of use with only a basic knowledge in microscopy and availability, confocal microscopy (CLSM and SDCM) can be considered as the actual gold standard for imaging, preferably fixed, 3D samples smaller than $\approx$ 100 μm.

6.2.4. *Photon reassignment approaches – the new confocal systems*

Lately, a new class of confocal microscopes has appeared making it possible to both benefit from the optical sectioning of conventional CLSM as well as to effectively take advantage of the spatial resolution an infinitesimally small pinhole would provide. It relies on the detection of the signal using a multipixel detector. For a position of the excitation focus in the sample, each pixel of the detector acts as a very small pinhole capturing a confocal image of the sample with an increased resolution according to its size: the smaller the pixel, the sharper the image, up to a limit defined by the Nyquist criterion for PSF sampling. The overlay of all these copies of the same area then provides a sharper and higher-intensity image as compared to the one a single (and larger) pinhole would capture. A complete image of the sample with enhanced resolution and intensity is finally built up summing the images for all the excitation focus positions throughout the sample similarly as in CLSM/SDCM.

This principle mainly relies on the accurate and precise overlay of the signal captured by each pixel of the detector for each excitation focus position, a process we refer to as photon re-assignement. This process considers the effective probability of detecting a fluorescence signal by each pixel of the 2D detector. On one side, the probability to excite a fluorophore is determined by the excitation's PSF (PSF_{exc}), which is centered onto the excitation focus position. On the other side, the probability of collecting a photon by a given pixel of the detector is characterized by its detection's PSF (PSF_{det}), which is centered onto the considered pixel. The probability of detecting a fluorescent signal on that pixel is hence given by the product of those two PSFs, which defines the effective detection's PSF (PSF_{eff}).

The PSF_{eff} maximum position then determines the location where the fluorescence signal most likely originates from when detected by that off-axis pixel. This maximum is actually shifted by a factor of m according to the PSF_{det} position (see Figure 6.3(A)). This factor is usually approximated by ½ which corresponds to the case of identical excitation and detection PSFs, i.e. when the fluorescence Stokes shift is neglected. The position of the photons detected by each pixel of the detector can thus be reassigned to a location about half the original distance from the excitation focus leading to a sharper image by a factor of ≈ ½ (see Figure 6.3(B)) (Sheppard 1988; McGregor et al. 2015).

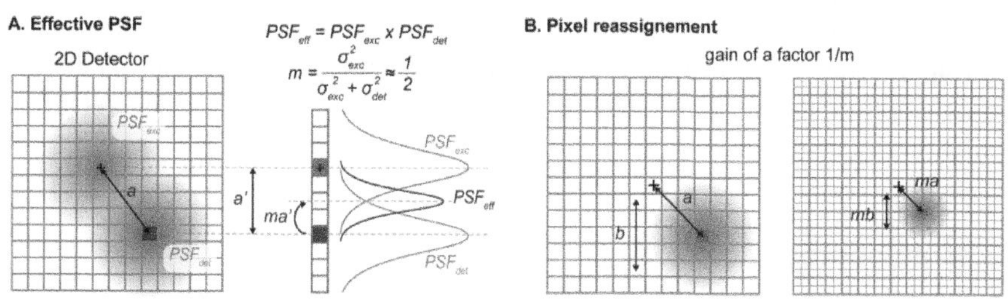

Figure 6.3. *A. Principle of confocal detection by an off-axis pixel. The effective PSF, which corresponds to the probability of detecting a fluorescent signal onto a given pixel, is defined by the product of the excitation and detection's PSFs. B. The location of PSF_{eff} maximum indicates that the signal collected by each pixel can be reassigned to m times its original distance (a) from the excitation spot position, resulting in a final image that is sharper by a factor of 1/m. For a color version of this figure, see www.iste.co.uk/sibarita/photonic.zip*

This strategy can be applied both using a single excitation focus (Müller and Enderlein 2010), or using several parallel foci for faster acquisition rates (York et al. 2012) similarly to SDCM. Photon reassignment can then be performed numerically, or in an all-optical fashion (Roth et al. 2013), enabling the imaging speed to be further increased up to 100 s Hz when multiple excitation foci are considered (York et al. 2013).

This acquisition approach provides several advantages as compared to standard confocal detection through a physical pinhole. It bypasses the practical limitations associated with the collection efficiency by a very small pinhole, effectively enabling the acquisition of confocal images with enhanced resolution. In addition, it better preserves the collected photons by re-assigning off-axis collected photons instead of being physically rejected. This leads to better collection efficiency, reducing the required illumination dose for the same image contrast, making this approach better suited for 3D live imaging. Nevertheless, the sample has to be probed four times more (pixels twice as small) to effectively benefit from this resolution gain, which somehow mitigates the effective illumination dose reduction. Since the out-of-focus signal is no longer physically filtered, it also imposes a limitation on the tolerable amount of background for which the in-focus signal remains discernible above the induced shot noise, thereby affecting the sample thickness that can be imaged.

It is also worth mentioning that confocal detection for 3D imaging can also be applied to label-free techniques such as reflectance microscopy. In this case, only the signal reflected from structures (reflective structures, particles, skin, retina, etc.) that are in the focal plane will create a tightly focalized spot at the pinhole plane. Such approaches have also been extended using photon reassignment strategy to further improve the spatial resolution of the sectioned imaged obtained (DuBose et al. 2019; Ta et al. 2023).

6.3. Multi-photon microscopy

6.3.1. *Principle*

The single photon-based absorption process commonly used in CLSM and SDCM is directly proportional to the illumination intensity. It thus occurs along the whole laser path through the sample generating both in and out-of-focus signals that need to be filtered to provide optical sectioning for 3D imaging (see Figures 6.2(A) and 6.4(A)). But nonlinear absorption processes that involve more than one photon and provide intrinsic 3D confinement of the excitation process also exist. They lead to completely new imaging properties that are particularly interesting for deep 3D imaging into living and opaque tissues.

The most widely used nonlinear absorption process in biological imaging is fluorescence excitation by two (or more) photons, called multi-photon microscopy (MPM). It occurs when at least two photons simultaneously reach a molecule and combine their energies to induce a molecular transition to an excited electronic state, which then proceeds along the normal fluorescence-emission pathway (see Figure 6.4(B)). In practice, the time window for the simultaneous absorption of two-photons by a fluorophore is less than a femtosecond, which is almost impossible under normal continuous illumination and requires concentrating excitation light both in space and in time. The possibility of such a multi-photon absorption process was predicted as early as 1931 by Maria Göppert-Mayer (1931). However, it was not until the 1990s, with the development of a pulsed laser with a pulse duration below 1 ps and high repetition rates ($\approx$ 100 MHz), that effective laser scanning MPM could be achieved. Indeed, in addition to spatial concentration through high NA objectives, such pulsed lasers have offered the possibility to temporally concentrate the photons within very short pulse duration with very high peak intensities. This combination has made it feasible to achieve sufficiently high photon densities within the focal spot of a laser to induce enough multi-photon absorption events for fluorescence imaging (Denk et al. 1990; Helmchen and Denk 2005).

In MPM, the signal I_{Fluo} generated depends nonlinearly on the photon density. It restricts the detectable fluorescence signal to the laser focal volume where the photon density is high enough to induce enough multi-photon absorption events. This leads to an intrinsic confinement of the excitation process, and therefore to optical sectioning, without having to spatially filter out-of-focus signal (see Figure 6.4(A)). In addition to this optical sectioning capability, it enables the use of all collected photons, including those that have undergone scattering events, to build up the final image. This same principle also works for the excitation side as scattered photons are generally too few to generate any signal. It makes MPM more robust against scattering properties of the sample and thus more suited for deep imaging (> 100 s µm) as compared to one-photon based imaging. As an example, it enabled imaging more than 1 mm in depth within opaque samples such as intact mouse brains (Horton et al. 2013; Ouzounov et al. 2017).

Another advantage of MPM enabling deep imaging in living tissues comes from the fact that the excitation light has a wavelength n-times longer than the corresponding excitation wavelength for a one-photon absorption process. For commonly used markers, excitation therefore occurs in the near-infrared spectral range (700–1300 nm), whereas emission arises in the visible spectral range.

In addition, in the near-infrared spectral range, light undergoes less absorption and scattering due to a lower absorption of light by water and hemoglobin, two main constituents of biological tissues, and scattering dependency on λ^{-4} (Rayleigh scattering). This spectral region is commonly called the biological optical window, which allows deeper imaging into biological tissues than in the visible range (see Figure 6.4(C)). In addition, excitation photons in MPM interact much less with the sample throughout their propagation, decreasing the probability of generating ROS from the excitation of any chromophores, making it more suitable for long-term live imaging. MPM is therefore the tool of choice for in vivo imaging within living and even behaving animals and is typically used for in vivo functional brain imaging. Nevertheless, in the focal plane, photobleaching and photo-toxic effects still occur due to multi-photon absorption.

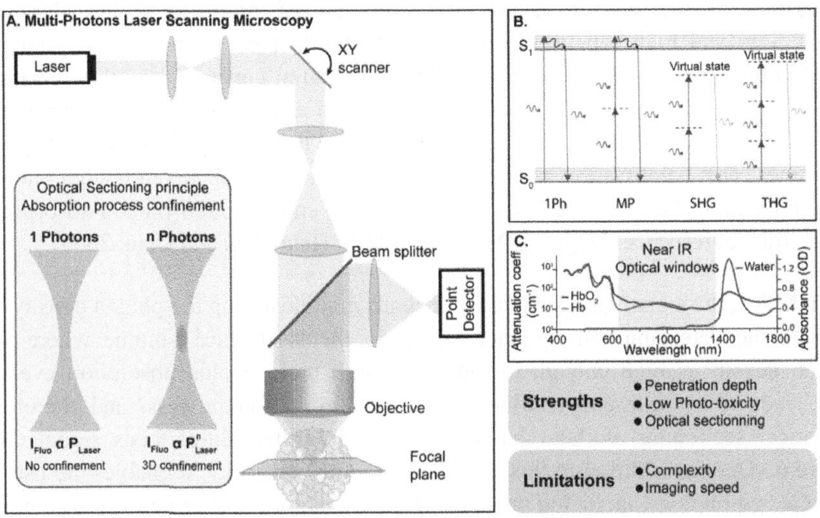

Figure 6.4. *A. MPM standard optical setup. The inset illustrates the optical sectioning principle arising from the photon density required to effectively induce absorption. B. Jablonsky diagram representing different absorption processes that can be used for biological imaging (1Ph: one photon; MP: multi-photons; SHG: second harmonic generation; THG: third harmonic generation). C. Biological optical windows defined as the spectral range for which the absorption and scattering of biological tissues are the lowest and therefore the imaging penetration depth the highest. For a color version of this figure, see www.iste.co.uk/sibarita/photonic.zip*

We can also briefly mention that other nonlinear multi-photon-based interactions with matter also exist that can provide possibly label-free 3D imaging capacities (see

Figure 6.4(B)). We can cite in particular second harmonic generations (SHG) and third harmonic generations (THG) (Campagnola et al. 2001; Weigelin et al. 2016), which enable probing specific matter organization or mismatch index interfaces respectively, and coherent anti-Stokes Raman scattering (CARS) (Krafft 2012), enabling us to probe specific chemical bonds.

6.3.2. *Laser-scanning multi-photon microscopy optical setup*

As MPM requires a very high density of photons, it was first made possible by combining both a concentration of photons in time, using a pulsed laser, and in space, focalizing the laser beam through a high NA objective. Most MPM architectures therefore rely on a point scanning approach, similarly to CLSM described previously, to build up an image. Combined with a pulsed laser exhibiting a very short pulse duration and repetition rate matching typical fluorescence lifetime, it generates sufficient multi-photon absorption for imaging. Key parameters to consider for the laser sources include the available power at the desired wavelength and the pulse duration. In that respect, the most widely used laser for MPM is the titanium–sapphire (Ti:sapphire) oscillator which displays a typical pulse duration of $\approx$ 100 fs and a repetition rate of approximately 100 MHz. In addition, its emission wavelength can be tuned in between 670 and 1070 nm, allowing the excitation of many common fluorophores used in biology. The pulsed laser is then scanned by a deflector module, usually composed of galvanometric mirrors, and then sent to the objective for focalization into the sample.

In detection, as compared to a confocal system, the excitation confinement means that we do not need to spatially filter the light through a pinhole anymore. Any collected photons, even the scattered ones, can contribute to the image formation without impairing the optical sectioning capacity but rather enhancing the imaging sensitivity. Thus, a simpler design can be used without requiring descanning the collected light in order to capture as many photons as possible, both ballistic and scattered (see Figure 6.4(A)), and prevent any photon loss through the optics.

6.3.3. *Widefield multi-photon microscopy by temporal focusing*

Even though the laser scanning MPM approach has proven to be extremely powerful for 3D live imaging in depth, it remains a scanning method with fundamentally limited temporal resolution. Similarly to CSLM, the imaging speed can be improved using faster resonant scanners or acousto-optics deflectors but their temporal resolution remains ultimately limited by the minimum pixel dwell time, enabling enough photons to be collected. This limitation was overcome in 2005 by the first demonstration of a widefield scanless temporal focusing (TF) approach, enabling an entire depth-resolved plane to be excited using a two-photon absorption process (Oron et al. 2005) (see Figure 6.5(A)).

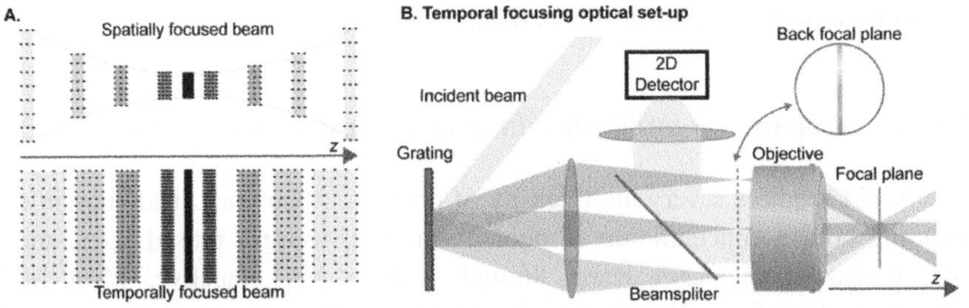

Figure 6.5. *A. Temporal focusing principle. Classically photons concentrated within short pulses are focused spatially by a high NA objective to achieve a sufficiently high photon density at the focal plane (top). In temporal focusing, each laser pulse is stretched in time and temporally focused by the objective lens leading to an increased photon density at the focal plane. B. Temporal focusing optical setup. The incident laser beam is temporally dispersed by a grating before being focused again through the objective lens. For a color version of this figure, see www.iste.co.uk/sibarita/photonic.zip*

In TF, the photons of a weakly focused beam are temporally focused by the objective lens leading to a high photon density at the objective's focal plane and a rapidly decreasing photon density at out-of-focus planes (see Figure 6.5(A)). It results in an axially confined excitation process, leading to an optical sectioning similar to point scanning MPM methods but over a large 2D area. This is commonly achieved by reflecting a high-intensity incoming pulsed laser on a grating to stretch it in time by spatially dispersing its spectral components. The laser spot on the grating is then re-imaged into the sample by the objective, making all the laser pulse's spectral components converge onto the focal plane (see Figure 6.5(B)). Hence, the photon density gradually increases up to the focal plane, resulting in the axial confinement of the multi-photon absorption process (Papagiakoumou et al. 2020). This has been made feasible mostly through the development of higher-intensity pulsed lasers enabling a sufficiently high photon density to be induced even onto an extended 2D area.

The improved temporal resolution of TF over conventional scanning point MPM makes this approach particularly suited for fast in vivo functional imaging requiring simultaneous signal recording of several (up to hundreds) cells (Schrödel et al. 2013; Prevedel et al. 2014, 2016). However, optimal TF imaging faces significant challenges to preserve (1) the photon density at the focal plane despite its spreading onto a larger excitation area, and (2) the imaging penetration depth which is affected by light scattering as compared to traditional point-scanning MPM approaches. Therefore, smaller illumination areas targeted to selected regions of interest are usually preferred to optimize the photon density and limit the scattering effect on the image contrast.

6.4. Light-sheet fluorescence microscopy

Confocal and multi-photon-based imaging both rely on the scanning of a single, or several, laser focus(es) throughout the samples to build up 2D depth-resolved sections within samples of interest. Despite their numerous advantages for 3D imaging, these methods still have limitations in terms of temporal resolution, field-of-view, and, for most of them, photo-toxicity. 3D imaging of a vast variety of samples, living or not, would thus greatly benefit from faster and gentler 3D imaging techniques. For about 15 years, the advent of light-sheet fluorescence microscopy (LSFM) has proposed new solutions towards these goals and is pushing the limits of 3D imaging of an increasing number of diverse and complex samples.

6.4.1. *Principles and advantages*

The first light-sheet (LS) microscope implementation was described as early as 1902 by Siedentopf and Zsigmondy as a new type of dark-field microscopy for the investigation of colloids in solution (Siedentopf and Zsigmondy 1902). But it was only in 2004 that the first modern architecture dedicated to fluorescence imaging of biological samples (LSFM) was proposed by Huisken et al. with the implementation of the so-called single plane illumination microscope (SPIM) (Huisken et al. 2004).

The main concept of LSFM is to dissociate the excitation from the detection path in order to provide intrinsic optical sectioning, while leveraging on the efficient and highly parallelized widefield excitation and detection schemes. In an LSFM configuration, a first optical path is used to create a thin sheet of light, which will excite only a thin section of the sample from the side.

Then, the generated fluorescence signal is collected perpendicular to this LS by a second optical path in a widefield detection manner (see Figure 6.6(A)). For 3D imaging, the LS and detection plane, or the sample itself, are displaced along the axial direction of the detection's objective to sequentially acquire 2D sections of the sample.

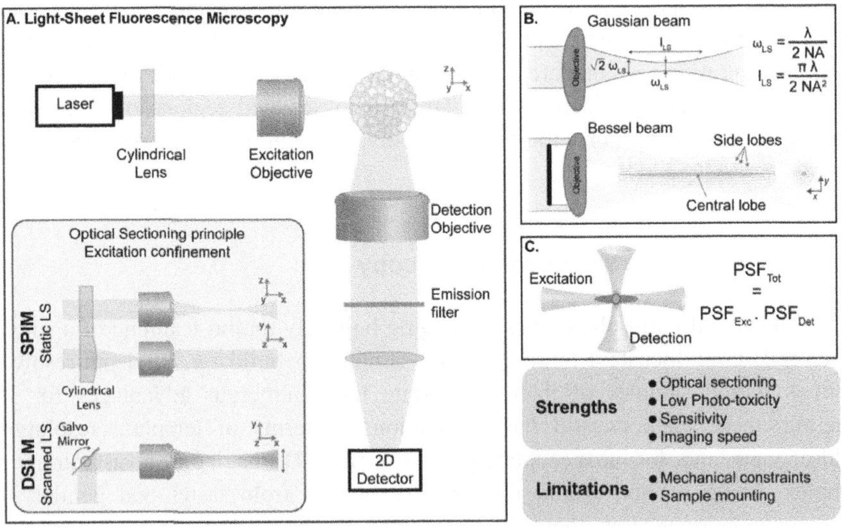

Figure 6.6. *A. LSFM basic architecture. The inset represents the two possible approaches to create an LS: a static LS through focalization by a cylindrical lens, or a virtually scanned one. B Diffraction laws make that a focused Gaussian beam rapidly diverges leading to a strong compromise in between the achievable LS thickness (ω_{LS}) and length (l_{LS}). An alternative to that trade-off is to use non-diffractive beams, such as Bessel beams, whose profile remains invariant along its propagation direction over much longer distances. C. The overall PSF of a LSFM is given by the product of the excitation and detection PSFs. For a color version of this figure, see www.iste.co.uk/sibarita/photonic.zip*

As opposed to epi-illumination approaches such as widefield microscopy, CLSM and MPM, this dissociated configuration considerably optimizes the use of the excitation light, as only the plane which is actually imaged is excited. Such optimization leads to several advantages for 3D imaging of living samples. First, the illumination of the sample from the side provides an intrinsic optical sectioning defined by the thickness of the LS. Second, the confinement of the excitation light to the imaged plane enables drastically reducing any unwanted photo-induced effects such as photobleaching and photo-toxicity. It makes LSFM one of the most adapted tools for the 3D imaging of living samples over a period of time that can last several hours to days with very little to no effects on their physiological developments. Third, widefield excitation and detection of a whole plane at once allows for very fast and sensitive imaging. It opens the possibility to monitoring rapid biological events and/or to further reducing the light dose received by the sample. And last but not least, the excitation and detection path dissociation provides endless possibilities for customizing the microscope arrangements (see Figure 6.7). It enables specifically adapting the system geometry to the sample constraints (size and morphology, accessibility, physiological development needs, environment control, etc.), and has thus opened up the possibility of imaging samples so far inaccessible due to mounting constraints.

The LSFM lateral resolution is the same as any standard widefield microscope and depends on the wavelength λ of the emitted light and the numerical aperture NA of the detection objective (see Figure 6.1(B)). Its axial resolution is, on its side, no longer determined solely by those two parameters, but depends as well on the LS thickness (see Figure 6.6(C)). It is therefore usually recommended to choose a detection objective having a greater DoF than the LS thickness. In that case, the system fully benefits from the optical sectioning ability provided by the LS thickness, as all the objects that are excited by the LS will appear in focus, and the detection PSF is reduced axially, improving its overall anisotropy. Conversely, if shorter DoF provides improved axial resolution, objects excited by the thicker LS may appear out-of-focus, potentially leading to a loss of the overall contrast.

6.4.2. *Light-sheet creation and properties*

Two main approaches exist to create an LS for LSFM. It can either be made through the focalization of a laser beam by a cylindrical lens as in the SPIM architecture initially proposed by Huisken et al. (2004), or by scanning a focalized laser beam perpendicularly to the detection objective to create a virtual LS as compared to the camera integration time (see Figure 6.6(A)). This later solution has led to the acronym DSLM for digitally scanned light-sheet microscopy (Keller et al. 2008). All LSFM are based on either the SPIM or DSLM architecture. Briefly, SPIM-based illumination enables us to better average the illumination intensity over the illumination area and frame rate as compared to DSLM approaches which create peak intensities along time and potentially induce higher photo-damages. However, SPIM methods are more prone to stripping artifacts arising when the excitation light encounters obstacles that scatter or absorb light along its propagation path into the sample. Yet, slightly rotating the LS around the detection objective's optical axis reduces those stripping artifacts as proposed in the mSPIM system (Huisken and Stainier 2007).

Most LSFMs use a Gaussian-shaped laser beam to create the LS. However, the diffraction of light dictates that such a beam diverges quickly away from its focus, leading to important trade-offs between the achievable LS thickness and its length. We usually consider the length of an LS (l_{ls}) as being the region for which the thickness of the beam is below $\sqrt{2}$ times its minimum thickness (ω_{ls}) (i.e. at the waist of the laser) (see Figure 6.6(B)). It follows that reducing the LS thickness leads to very short LS and consequently imaging field-of-views (FOVs), which become unusable in practice. On the contrary, increasing the length of the LS to achieve very large FOVs can be done but at the expense of its thickness and, by extension, its optical sectioning ability and axial resolution gain. To bypass this trade-off and allow imaging over wider FOVs, while preserving the axial resolution and sectioning, two main approaches have been proposed:

– A first method consists of rapidly sweeping a very thin, but short, Gaussian-based LS along the axial direction of the excitation objective. Then, tilling the portion of the sample illuminated with the thinnest part of the LS enables us to create a virtually very long and thin LS (Gao 2015). Such approaches offer the capacity to image very wide, centimeter scale, FOVs with quasi-isotropic and sub-cellular resolutions, but at the expense of a slower acquisition speed and sample over-exposition (Voigt et al. 2019).

– A second method consists of implementing, with a DSLM approach, non-diffractive beams, such as Bessel (Planchon et al. 2011) or Airy (Vettenburg et al. 2014) beams, whose profiles have the particularity of being invariant over much larger distances than Gaussian beams of comparable thickness (see Figure 6.6(B)). They also make it possible to further reduce stripes artifacts due to their self-reconstructing properties after encountering an obstacle. However, they usually come with important side lobes that degrade their sectioning ability and thus the imaging contrast. They are thus usually combined with multi-photon absorption processes to get rid of those less energetic side lobes (Fahrbach et al. 2013a), or with a confocal line detection scheme that filters out the fluorescence signal excited by those side lobes (Fahrbach and Rohrbach 2012). Finally, optical lattices produced by the interference of several well-defined plane-waves enable us to create very thin LSs over long distances with very little side lobes, and even lower illumination dose, in a SPIM-like configuration (Chen et al. 2014).

6.4.3. *Main implementation overview*

The uncoupling of the excitation and detection pathways in LSFM offers endless customization possibilities of the microscope arrangements. The microscope can now be specifically built around the samples to fulfill its specific physiological requirements, and samples can be embedded, suspended, enclosed into dedicated chambers or simply mounted onto a coverslip. Combined with its very low photo-toxicity, LSFM has opened the possibility to image samples and organisms that are growing, moving and even behaving, and explore new samples that were too fragile or complex to mount onto standard supports of more conventional imaging approaches (Wan et al. 2019). Yet, the large diversity of samples that can be imaged by LSFM has led to a plethora of different and dedicated architectures, and associated acronyms, that are sometimes difficult to navigate into. Here, we present some of the most common architectures used in the LSFM community (see Figure 6.7).

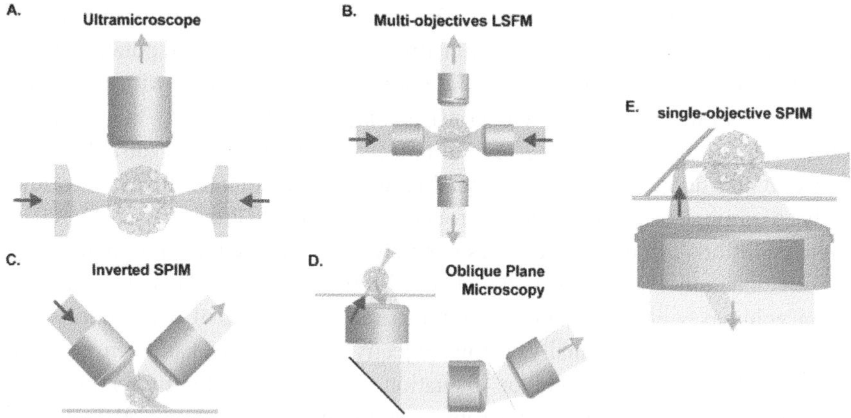

Figure 6.7. *Main LSFM implementations overview. For a color version of this figure, see www.iste.co.uk/sibarita/photonic.zip*

6.4.3.1. *Ultramicroscope*

A first category of LSFM, that we can gather in the so-called ultramicroscope domain, has been designed to accommodate 3D imaging of very large samples going from hundreds of micrometers up to centimeters such as organoids, whole embryos, brains or organs (Dodt et al. 2007; Belle et al. 2017). They rely on an upright configuration and usually use dual-sided illumination strategies based on multi-oriented static LS (SPIM mode) to increase the achievable FOV and reduce the stripes artifacts (Figure 6.7A). Combined with thin and axially swept LS, such systems can lead to quasi-isotropic sub-cellular resolution over centimeter scale FOV (Voigt et al. 2019).

6.4.3.2. *Horizontal multi-objective architectures*

The second, and largest, category of LSFMs are dedicated to image samples in the range of a few hundreds of micrometers encompassing a large variety of 3D cell cultures (spheroids, oncospheres, organoids, etc.) or small transparent organisms such as drosophila, C. elegans worms or zebrafishes. They are usually developed around the DSLM principle, using from two objectives up to four objectives. Their configurations are mostly horizontal, and the samples are mounted (hanged, suspended or embedded) onto specific chambers in between the objectives (see Figure 6.7(B)). They are particularly suited for developmental biology as they enable monitoring numerous sample developments for very long periods of time with virtually no photo-damaging effects.

These architectures greatly benefit from multi-view imaging approaches that involve imaging the same sample from different angles and computationally, or optically, fusing them to reconstruct the entire sample's volume with the highest spatial resolution. A first solution consists of physically turning the sample in between the two excitation and detection objectives (Swoger et al. 2007) but at the expense of an important acquisition speed reduction and complex fusing algorithms usually requiring beads embedded around the sample (Preibisch et al. 2010). Another solution is to use a three- (respectively, four-) objective configuration, two for excitation and one (respectively, two) for the detection. This multiplies the imaged volumes without impairing the temporal resolution and the computational cost (Krzic et al. 2012). Finally, a four-objective architecture has been proposed, with each objective used both for excitation and detection to further fasten the acquisition process and achieve isotropic resolution (Chhetri et al. 2015). If those implementations can provide impressive FOV and resolution, they may also be expensive and complex to set up. It is important to be aware that these solutions require storage capacity analysis and processing pipelines, and visualization tools at least as complexes, which potentially represents a major bottleneck.

6.4.3.3. *Inverted configurations*

To accommodate for more conventional sample mounting onto coverslips, a vertical arrangement of the illumination and detection pathways has been proposed with the (d)iSPIM (Wu et al. 2011) (see Figure 6.7(C)). These configurations are perfectly adapted to image samples spread or lying onto a coverslip. Additionally, by using each objective for excitation and detection alternatively, it enables multi-view imaging leading to isotropic resolution (Kumar et al. 2014). Combined with custom excitation objective and LS shaping, this architecture enables us to achieve single molecule localization super-resolution microscopy (SMLM) over whole cells as well as impressive imaging speed at very high spatial resolution (Chen et al. 2014).

Those architectures are perfectly suited to image samples ranging from single cells up to small and transparent organisms (*C. elegans*, zebrafishes, etc.) or cellular aggregates of hundreds of micrometers. In addition, it offers the possibility to image tissue sections such as brain slices that can be very complex to mount vertically in between two, or more, objectives.

6.4.3.4. *Single-objective configurations*

All the aforementioned architectures rely on the use of several objectives arranged perpendicularly from one another. Even if it opens the possibility to many implementations, it comes with some mechanical constraints. It first makes the mounting of the sample in between the objectives positioned very close to one another relatively complex. This limits the possibility of exploring a large area and/or restricts the overall throughput to one, or a few, samples at a time. Second, it prevents the possibility of using high NA objectives, which are bulkier with smaller working distances, and thus cannot fit into this perpendicular arrangement, somehow limiting the sensitivity and achievable spatial resolution.

To circumvent this mechanical limitation, it has been proposed to use a unique objective to both create the LS and collect the fluorescence signal. It can be achieved using 45° mirrors to reflect an incoming laser beam and create an LS perpendicular to the detection objective optical axis either using two oppositely positioned objectives (Gebhardt et al. 2013) or using a single objective (Galland et al. 2015) (see Figure 6.7(E)). High NA objective can then be used, opening the possibility to perform single molecule localization microscopy (SMLM) tens of microns deep as compared to standard SMLM approaches limited to imaging the first micron above the coverslip. In addition, the soSPIM technology (Galland et al. 2015), which uses dedicated microfabricated devices displaying 45° mirrors, offers an interesting alternative to increase LSFM imaging throughput.

Indeed, these microfabricated supports contain hundreds to thousands of micro-niches to culture and make grown 3D cell cultures and are compatible with 3D LSFM-based imaging. This makes them an ideal tool to standardize and parallelize both the culture and the imaging of these samples.

Another strategy for single-objective LSFM architecture is to create a tilted LS through the detection objective, and project the tilted excited plane onto a 2D camera using a combination of two objectives forming an angle the one from the other (Dunsby 2008) (see Figure 6.7(D)). This architecture offers a very high imaging speed over wide FOV (Voleti et al. 2019) and is compatible with any standard support such as coverslip and multi-well plates (Yang et al. 2019). It therefore opens new avenues to increase LSFM throughput and screening assay on 3D samples (Maioli et al. 2016; Yang et al. 2019). However, it comes with a slight loss of detection NA, which can be mitigated by adequately choosing the different objectives used (Yang et al. 2019), to allow in-depth SMLM imaging (Kim et al. 2019).

6.4.3.5. *Custom designs*

To circumvent the imaging throughput limitation of conventional LSFM architectures, or to enable the use of high NA objectives in detection, alternative custom designs still relying on multiple objectives have been proposed:

– An interesting solution to improve the throughput of standard LSFM architectures is to combine them with (micro-)fluidics approaches in order to sequentially flow multiple samples within the imaging chamber (Gualda et al. 2015; Paiè et al. 2016). This enables screening a large amount of sample very rapidly but is, however, not compatible with the monitoring of the same samples along time for developmental studies.

– To allow for high NA objective in detection, several teams have proposed various solutions whose main concept is to increase the angle in between the excitation and detection objectives. This can be achieved by deviating the LS propagation direction thanks to prisms (Hu et al. 2013; Hu et al. 2016), mirrors (Gebhardt et al. 2013; Zhao et al. 2014; Greiss et al. 2016), or by using slightly thicker and oblique LS according to the detection objective optical axis (Gustavsson et al. 2018; Ponjavic et al. 2018). Leveraging on the in-depth optical sectioning provided by the LS excitation scheme and the use of high NA objectives, those architectures offer the possibility to perform SMLM at the scale of whole cells.

6.5. Computational-based methods

To provide optical sectioning capacity for 3D imaging, all the aforementioned methods either manage to not generate out-of-focus signals by confining the excitation light (section 6.3 on MPM, or section 6.4 on LSFM), or rely on the physical filtering in the detection path of the out-of-focus signal that has been generated (section 6.2 on CLSM). A fourth strategy, which involves image post-processing, is, however, possible. Based on the conventional epi-illumination approach, those methods rely on the capacity to computationally identify and extract the signal that is in-focus from the overall signal detected (Mertz 2011).

Let us consider the excitation of the sample with a periodic light pattern, such as a sin wave or a grid pattern, in an epi-configuration. The axial broadening of the microscope PSF makes it so that this illumination pattern will be discernible only for the in-focus plane, whereas it will become increasingly blurry, and therefore homogeneous, for out-of-focus planes. The contrast of the pattern variation can then serve as an indicator of how much the sample is in-focus. By identifying only the signal which carries this contrast variations from the overall detected signal, it becomes possible to extract the in-focus components and provide an optically sectioned image. This is the principle of *Structured Illumination Microscopy* (SIM) as proposed in 1997 by Neil et al. (1997).

In a practical point of view, SIM requires the acquisition of a sequence of images while translating the periodic pattern in one direction stepwise. Onto the camera, the in-focus signal will therefore be modulated in regard to this illumination pattern translation at the contrary of the out-of-focus signal components. A demodulation algorithm (Neil et al. 1997) can then infer the variations of the illumination and therefore their contrast.

Eventually, an optically sectioned image can be reconstructed from this inferred contrast at each camera's pixel. To be effective, each position of the in-focus plane needs to be eventually homogeneously illuminated, requiring a minimum of three images to be acquired, somehow limiting the achievable imaging speed. It is worth noting that the excitation of the sample with a structured illumination not only provides a contrast variation on the in-focus plane but also adds additional spatial frequencies to the emitter's distribution in the form of Moiré fringes. Those spatial frequencies can be used to extract additional information and improve the achievable spatial resolution beyond the diffraction limit as proposed by Gustafsson (2000, 2005) and Gustafsson et al. (2008) and described in Chapter 7 using more advanced reconstruction algorithms.

To further increase the temporal resolution of SIM and overcome the constraint to precisely move a grid pattern onto the sample, a simpler strategy called HiLo microscopy has been proposed (Lim et al. 2008). HiLo microscopy only requires the acquisition of one structured image and one conventional homogeneously illuminated image. The fusion of the high-resolution information of the homogeneously illuminated image with the low-resolution information of the demodulated patterned image enables the reconstruction of a fully-resolved optically sectioned image. HiLo microscopy has the advantage of providing a 3D optically sectioned image at a video high frame rate for fast dynamic recording. It is also insensitive to sample-induced distortion of the illumination pattern, making it more robust for deeper imaging as compared to conventional SIM approaches, up to the possibility of using random illumination speckle for the patterned illumination (Lim et al. 2011).

SIM and HiLo microscopy approaches enable providing optically sectioned images with somehow pretty simple equipment and at a low cost as compared to other approaches such as CSLM or MPM. However, their performances depend on the in-depth contrast variations of the illumination pattern, the finite camera dynamics and the shot noise induced by the out-of-focus signal. It sets a strong limit in the sample thickness, scattering and brightness that can be imaged for the modulated in-focus components to be discernible above the background signal and associated noise. A solution to this problem is to combine the principle of SIM background rejection with the confinement of the excitation light provided by LS illumination schemes to further improve LSFM optical sectioning (Kalchmair et al. 2010; Keller et al. 2010; Mertz and Kim 2010).

6.6. Futures challenges for 3D imaging of complex samples

As discussed in this chapter, four main families of methods have been developed providing in-depth optical sectioning to assess the third dimension and observe biological samples in their entire volume.

However, imaging always bigger and more complex samples, monitoring always faster functional events and extracting more content from the 3D images still raise many challenges in several domains:

– How can we improve the effective penetration depth that can be achieved with sufficient spatial resolution to probe proteins' organization and/or dynamics within always more complex and physiologically relevant samples?

– Can we increase the imaging speed to more faithfully monitor fast dynamical events in 3D?

– How can we handle, manage and analyze always growing datasets in 3D to allow the representation and the extraction of quantitative and reliable information?

6.6.1. *Optical aberrations*

Microscopes rely heavily on high-quality optical systems to obtain the optimum resolution determined by the diffraction limit of light (see Figure 6.1(B)). However, even when considering perfect optical components, the imaging performances are affected by the optical properties of the specimen being studied. Absorption and scattering limit the achievable penetration depth, whereas sample inhomogeneities introduce optical aberrations which become increasingly dramatic when imaging deeper into the sample (see Figure 6.8(A) and (B)). This results in a loss of signal and contrast as well as the distortion of the system's PSF both laterally and axially, eventually impacting the resolution and optical sectioning capabilities of the system. We therefore have to compensate for and/or suppress those aberrations to image at high spatial resolution deep into complex samples.

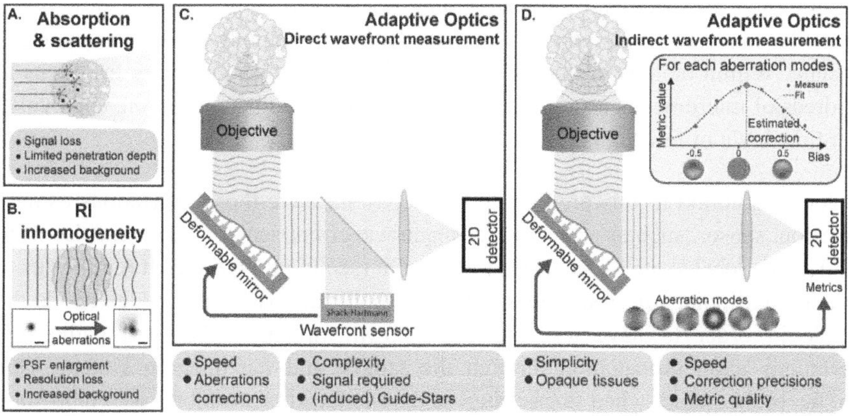

Figure 6.8. *Impact of absorption and scattering (A) and sample inhomogeneities (B) on light propagation and image formation. (C) Direct wavefront measurement and (D) indirect sensor-less wavefront measurement principle for optical aberrations correction via AO. For a color version of this figure, see www.iste.co.uk/sibarita/photonic.zip*

A first static solution for imaging deep into thick samples is to make the samples more transparent to light through the use of clearing approaches (Ueda et al. 2020). Using chemically based approaches, clarification methods aim to remove the pigments of tissues that are responsible for light absorption, and homogenize their index of refraction to lower optical aberrations and light scattering. These methods make it possible to image very large, and even previously opaque samples (Belle et al. 2017), but are fixed, which prevents probing the dynamic phenomenon.

Another way to correct for those aberrations and make it possible to image deeper into samples is to directly correct for them by compensating their effects on the light wavefront thanks to an active component (Ji 2017). This solution, called adaptive optics (AO), is derived from Astronomy, where the aim is to overcome the effect of atmosphere turbulences. Its principle is pretty straightforward: if the aberrations to be corrected can be assessed, a wavefront modulator can induce a wavefront distortion inverse to those aberrations and minimize, or even cancel, their effects on the imaging performances. Wavefront modulators such as deformable mirror (DM) or spatial light modulator (SLM), enable us to quite easily and robustly shape the phase of any optical wave. The challenge of any AO system comes, however, from the way the optical aberrations can be assessed. In addition, optical aberrations that mainly arise from sample inhomogeneities, may rapidly evolve both in space and in time. Their correction therefore implies the capacity to regularly measure them in space and time according to so-called isoplanatic patches, which define the spatio-temporal volumes over which the aberrations can be considered as constant. Within biological tissues, the size of those patches may vary from tens to hundreds of micrometers according to their complexity and composition, and evolve over time scale of few hours.

A first solution to this problem is to directly measure those aberrations using a wavefront sensor, such as a Shack–Hartmann wavefront sensor, which enables us to evaluate the local slope of the probed wavefront (see Figure 6.8(C)). It requires us to have isolated point sources, at least one per isoplanatic patch, embedded inside the sample in order to have an object of known wavefront that will gather the aberrations upon propagation through the sample and the imaging system. This can be problematic when measuring sample-induced aberrations as embedding fiduciaries into the sample can be cumbersome or even toxic and/or impossible. To bypass this problem, several works have proposed to rely on nonlinear absorption processes in order to artificially produce a localized point source directly inside the sample (Aviles-Espinosa et al. 2011). It, however, complexifies the optical setup, and only works when sufficient ballistic, unscaterred, photons can be collected and sent towards the wavefront sensor. As a result, direct wavefront sensing is most of the time limited to image fairly transparent tissues or restricted to image the superficial layers of opaque ones.

A second approach is to indirectly estimate the aberrations from a metric (brightness, contrast, spatial frequency content, etc.) that is computed from the microscope image/signal itself. In that case, the wavefront modulator is iteratively adjusted until this metric is optimized. Appropriate selection of this metric is therefore crucial and depends on the sample and/or imaging modality used. As a general rule, image-brightness-based metrics are usually well adapted for bright and stable samples, whereas metrics assessing the spatial frequency content, or contrast of the image, are more suitable for dimmer and/or fluctuating signals. A convergence algorithm is then used to compute the wavefront shape that will maximize the metric value. Nowadays, the modal approach that consists of sequentially optimizing each mode of the decomposition of aberrations along an orthogonal basis (most commonly along the Zernike polynomials basis) is preferred because it substantially reduces the number of images that need to be acquired (Booth 2014) (see Figure 6.8(D)).

Direct and indirect sensing propose two different solutions to correct for optical aberrations. If direct sensing is the fastest and more precise method, it requires a higher system complexity and the capacity to have bright point-emitters inside the sample, drastically limiting its possible use. On the other side, indirect sensing is simpler to implement and can be applied to opaque samples, but they are slower (seconds to minutes) in measuring aberrations and may not correct for all of them. In any case, both approaches in correcting system- and sample-induced aberrations help optimizing the collection of photons and the contrast of the image, thus increasing the penetration depth of the imaging method used.

6.6.2. *Fast 3D imaging methods*

3D imaging in fluorescence microscopy mostly relies on the sequential acquisition of multiple 2D planes, which, after combination, enables us to reconstruct a volumetric image of the sample studied (see Figure 6.1(D)). This mainly comes from the fact that the microscope objective's DoF is highly confined in the axial direction (see Figure 6.1(C)). Therefore, there is a need to axially scan both the excitation plane and the objective focal plane throughout the sample to produce a sharp image of its entire volume. However, this sequential scanning process strongly limits the achievable acquisition speed and prevents the possibility to observe very fast biological events. Several attempts have therefore been made to increase this 3D imaging speed:

– This can be achieved by increasing the objective's DoF through the shaping of the objective PSF. Instead of using sharp PSF in the axial direction, the concept is to use axially elongated PSF to project all 3D information of an extended volume into a 2D plane at once (Dowski and Cathey 1995) (see Figure 6.9(A)). If those methods have successfully been applied to study the fast functional activities of neurons in the living brain (Quirin et al. 2014), they come at the expense of axial information and cannot be applied to dense samples under the penalty of losing or mixing information lying in the same axial direction.

– Increasing volumetric imaging speed can also be achieved through remote focusing approaches. An important limitation for fast 3D imaging is the need to move either the detection objective or the sample itself. But the weight of the objectives limits the speed at which it is possible to move them, and moving too fast the sample may impact its behaving and viability. The possibility to adjust the position of the objective focal plane remotely without moving either the objective or the sample can thus lead to very high imaging speed. This can be achieved by conjugating either a movable mirror to the image plane (Botcherby et al. 2012) or an electrically tunable lens (ETL), a spatial light modulator or a deformable mirror to the objective's back focal plane (Fahrbach et al. 2013b) (see Figure 6.9(B)), leading to volumetric imaging at a rate of up to 100 s of Hz. It thus makes it possible to faithfully monitor very fast 3D processes such as the beating heart of small animals.

– Another solution to increase the imaging speed of 3D volumes is to simultaneously detect and record the signal of several different focal planes slightly defocused one from the other inside the sample (see Figure 6.9(C)). Thanks to a dedicated diffractive optical element positioned in the detection path of the microscope, multi-focus microscopy approach enables to simultaneously image up to 25 different focal planes in a single camera frame leading to very fast volumetric imaging. However, it comes at the expense of the field of view, a complex optical setup and a photon budget spread between each sub-image (Abrahamsson et al. 2013).

– Finally, light-field approaches also propose a solution to image an entire volume in a single camera frame for very high volumetric imaging speed (Prevedel et al. 2014) but at the expense of lower spatial resolution and field of views. It uses a microlens array positioned at the image plane to capture both the position and angle of the incident light in a single frame. It then relies on important processing steps to reconstruct the 3D content of the imaged volume.

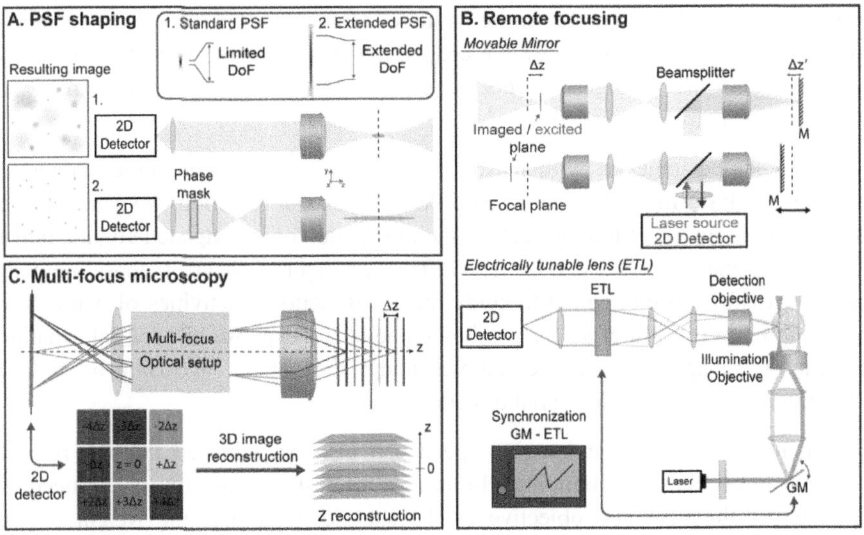

Figure 6.9. *Fast 3D acquisition. A. PSF shaping approach: the PSF is axially extended thanks to a phase mask in the detection path in order to increase the microscope depth of field and thus produce a net image of an axially extended volume in a single frame. B. Remote focusing approaches: the imaged (respectively, excided) plane is axially adjusted remotely thanks either to a movable mirror conjugated to the objective focal plane (top) or to an electrically tunable lens conjugated to the objective's back focal plane (bottom). C. In multi-focus microscopy, several slightly defocused (Δz) planes are simultaneously acquired and projected onto a different area of a single camera (respectively, different cameras). For a color version of this figure, see www.iste.co.uk/sibarita/photonic.zip*

6.6.3. *Data management, analysis and visualization*

A last challenge, when we aim to perform 3D imaging, is the huge amount of data that can be easily generated, which have to be stored, visualized and analyzed to extract relevant and quantitative contents from them. It is particularly striking when the imaging method enables imaging a living sample along a prolonged period of time, such as LSFM techniques. It is therefore important to anticipate such requirements to prevent this from becoming the main limiting factor of the acquisition pipeline, regardless of the imaging method used.

6.7. References

Abrahamsson, S. et al. (2013). Fast multicolor 3d imaging using aberration-corrected multifocus microscopy. *Nature Methods*, 10(1), 60–63. doi: 10.1038/nMeth.2277.

Aviles-Espinosa, R. et al. (2011). Measurement and correction of in vivo sample aberrations employing a nonlinear guide-star in two-photon excited fluorescence microscopy. *Biomedical Optics Express*, 2(11), 3135–3149. doi: 10.1364/BOE.2.003135.

Belle, M. et al. (2017). Tridimensional visualization and analysis of early human development. *Cell*, 169, 161–173. doi: 10.1016/j.cell.2017.03.008.

Booth, M. (2014). Adaptive optical microscopy: The ongoing quest for a perfect image. *Light: Science & Applications*, 3(4), e165. doi: 10.1038/lsa.2014.46.

Botcherby, E.J. et al. (2012). Aberration-free three-dimensional multiphoton imaging of neuronal activity at kHz rates. *Proceedings of the National Academy of Sciences of the USA*, 109(8), 2919–2924. doi: 10.1073/pnas.1111662109.

Campagnola, P.J. et al. (2001). Second-harmonic imaging microscopy of living cells. *Journal of Biomedical Optics*, 6(3), 277. doi: 10.1117/1.1383294.

Chen, B.C. et al. (2014). Lattice light-sheet microscopy: Imaging molecules to embryos at high spatiotemporal resolution. *Science*, 346(6208), 1257998. doi: 10.1126/science.1257998.

Chhetri, R.K. et al. (2015). Whole-animal functional and developmental imaging with isotropic spatial resolution. *Nature Methods*, 12(12), 1171–1178. doi: 10.1038/nmeth.3632.

Conchello, J.A. and Lichtman, J.W. (2005). Optical sectioning microscopy. *Nature Methods*, 2(12), 920–931. doi: 10.1038/nmeth815.

Denk, W. et al. (1990). Two-photon laser scanning fluorescence microscopy. *Science*, 248, 73–76 [Online]. Available at: http://www.pubmedcentral.nih.gov/articlerender.fcgi?artid=3481855&tool=pmcentrez&rendertype=abstract.

Dodt, H. et al. (2007). Ultramicroscopy: Three-dimensional visualization of neuronal networks in the whole mouse brain. *Nature Methods*, 4(4), 331–336. doi: 10.1038/NMETH1036.

Dowski, E.R. and Cathey, W.T. (1995). Extended depth of field through wave-front coding. *Applied Optics*, 34(11), 1859. doi: 10.1364/ao.34.001859.

DuBose, T.B. et al. (2019). Super-resolution retinal imaging using optically reassigned scanning laser ophthalmoscopy. *Nature Photonics*, 13(4), 257–262. doi: 10.1038/s41566-019-0369-7.

Dunsby, C. (2008). Optically sectioned imaging by oblique plane microscopy. *Optics Express*, 16(25), 20306–20316 [Online]. Available at: http://www.ncbi.nlm.nih.gov/pubmed/19065169.

Fahrbach, F.O. and Rohrbach, A. (2012). Propagation stability of self-reconstructing Bessel beams enables contrast-enhanced imaging in thick media. *Nature Communications*, 3, 632. doi: 10.1038/ncomms1646.

Fahrbach, F.O. et al. (2013a). Light-sheet microscopy in thick media using scanned Bessel beams and two-photon fluorescence excitation. *Optics Express*, 21(11), 13824. doi: 10.1364/oe.21.013824.

Fahrbach, F.O. et al. (2013b). Rapid 3D light-sheet microscopy with a tunable lens. *Optics Express*, 21(18), 21010. doi: 10.1364/oe.21.021010.

Galland, R. et al. (2015). 3D high- and super-resolution imaging using single-objective SPIM. *Nature Methods*, 12(7), 641–644. doi: 10.1038/nmeth.3402.

Gao, L. (2015). Extend the field of view of selective plan illumination microscopy by tiling the excitation light sheet. *Optics Express*, 23(5), 6102. doi: 10.1364/OE.23.006102.

Gebhardt, J.C.M. et al. (2013). Single-molecule imaging of transcription factor binding to DNA in live mammalian cells. *Nature Methods*, 10, 421–426. doi: 10.1038/nmeth.2411.

Giepmans, B.N.G. et al. (2006). The fluorescent toolbox for assessing protein location and function. *Science (New York)*, 312(5771), 217–224. doi: 10.1126/science.1124618.

Göppert-Mayer, M. (1931). Elementary processes with two quantum transitions. *Annalen der Physik (Leipzig)*, 9(273). doi: 10.1002/andp.200910358.

Greiss, F. et al. (2016). Single-molecule imaging in living drosophila embryos with reflected light-sheet microscopy. *Biophysical Journal*, 110(4), 939–946. doi: 10.1016/j.bpj.2015.12.035.

Gualda, E.J. et al. (2015). SPIM-fluid: Open source light-sheet based platform for high-throughput imaging. *Biomedical Optics Express*, 6(11), 4447. doi: 10.1364/BOE.6.004447.

Gustafsson, M.G.L. (2000). Surpassing the lateral resolution limit by a factor of two using structured illumination microscopy. *Journal of Microscopy*, 198(Pt 2), 82–87. doi: 10.1046/j.1365-2818.2000.00710.x.

Gustafsson, M.G.L. (2005). Nonlinear structured-illumination microscopy: Wide-field fluorescence imaging with theoretically unlimited resolution. *Proceedings of the National Academy of Sciences of the USA*, 102(37), 13081– 13086. doi: 10.1073/pnas.0406877102.

Gustafsson, M.G.L. et al. (2008). Three-dimensional resolution doubling in wide-field fluorescence microscopy by structured illumination. *Biophysical Journal*, 94(12), 4957–4970. doi: 10.1529/biophysj.107.120345.

Gustavsson, A.K. et al. (2018). 3D single-molecule super-resolution microscopy with a tilted light sheet. *Nature Communications*, 9(123). doi: 10.1101/135699.

Helmchen, F. and Denk, W. (2005). Deep tissue two-photon microscopy. *Nature Methods*, 2(12), 932–940. doi: 10.1038/nmeth818.

Horton, N.G. et al. (2013). In vivo three-photon microscopy of subcortical structures within an intact mouse brain. *Nature Photonics*, 7(3), 205–209. doi: 10.1038/nphoton.2012.336.

Hu, Y.S. et al. (2013). Light-sheet Bayesian microscopy enables deep-cell super-resolution imaging of heterochromatin in live human embryonic stem cells. *Optical Nanoscopy*, 2(1), 7. doi: 10.1186/2192-2853-2-7.

Hu, Y.S. et al. (2016). Superresolution imaging reveals nanometer- and micrometer-scale spatial distributions of T-cell receptors in lymph nodes. *Proceedings of the National Academy of Sciences of the USA*, 113(26), 7201–7206. doi: 10.1073/pnas.1512331113.

Huisken, J. and Stainier, D.Y.R. (2007). Even fluorescence excitation by multidirectional selective plane illumination microscopy (mSPIM). *Optics Letters*, 32(17), 2608. doi: 10.1364/ol.32.002608.

Huisken, J. et al. (2004). Optical sectioning deep inside live embryos by selective plane illumination microscopy. *Science (New York)*, 305(5686), 1007–1009. doi: 10.1126/science.1100035.

Ji, N. (2017). Adaptive optical fluorescence microscopy. *Nature Methods*, 14(4), 374–380. doi: 10.1038/nmeth.4218.

Kalchmair, S. et al. (2010). Image contrast enhancement in confocal ultramicroscopy. *Optics Letters*, 35(1), 79. doi: 10.1364/ol.35.000079.

Keller, P.J. et al. (2008). Reconstruction of zebrafish early embryonic development by scanned light sheet microscopy. *Science*, 322(5904), 1065–1069. doi: 10.1126/science.1162493.

Keller, P.J. et al. (2010). Fast, high-contrast imaging of animal development with scanned light sheet-based structured-illumination microscopy. *Nature Methods*, 7(8), 637–642. doi: 10.1038/nmeth.1476.

Kim, J. et al. (2019). Oblique-plane single-molecule localization microscopy for tissues and small intact animals. *Nature Methods*, 16, 853–857. doi: 10.1038/s41592-019-0510-z.

Krafft, C. (2012). Raman and coherent anti-stokes Raman scattering microspectroscopy for biomedical applications. *Journal of Biomedical Optics*, 17(4), 040801. doi: 10.1117/1.jbo.17.4.040801.

Krzic, U. et al. (2012). Multiview light sheet microscope for rapid in toto imaging. *Nature Methods*, 9(7), 730. doi: 10.1038/Nmeth.2064.

Kumar, A. et al. (2014). Dual-view plane illumination microscopy for rapid and spatially isotropic imaging. *Nature Protocols*, 9(11), 2555–2573. doi: 10.1038/nprot.2014.172.

Lim, D. et al. (2008). Wide-field fluorescence sectioning with hybrid speckle and uniform-illumination microscopy. *Optics Letters*, 33(16), 1819. doi: 10.1364/ol.33.001819.

Lim, D. et al. (2011). Optically sectioned in vivo imaging with speckle illumination HiLo microscopy. *Journal of Biomedical Optics*, 16(1), 1. doi: 10.1117/1.3528656.

Maioli, V. et al. (2016). Time-lapse 3-D measurements of a glucose biosensor in multicellular spheroids by light sheet fluorescence microscopy in commercial 96-well plates. *Scientific Reports*, 6(November), 37777. doi: 10.1038/srep37777.

McGregor, J.E. et al. (2015). Post-processing strategies in image scanning microscopy. *Methods*, 88, 28–36. doi: 10.1016/j.ymeth.2015.05.002.

Mertz, J. (2011). Optical sectioning microscopy with planar or structured illumination. *Nature Methods*, 8(10), 811–819. doi: 10.1038/nmeth.1709.

Mertz, J. and Kim, J. (2010). Scanning light-sheet microscopy in the whole mouse brain with HiLo background rejection. *Journal of Biomedical Optics*, 15(01), 1. doi: 10.1117/1.3324890.

Müller, C.B. and Enderlein, J. (2010). Image scanning microscopy. *Physical Review Letters*, 104(19), 198101. doi: 10.1103/PhysRevLett.104.198101.

Neil, M.A.A. et al. (1997). Method of obtaining optical sectioning by using structured light in a conventional microscope. *Optics Letters*, 22(24), 1905. doi: 10.1364/ol.22.001905.

Oron, D. et al. (2005). Scanningless depth-resolved microscopy. *Optics Express*, 13(5), 1468–1476.

Ouzounov, D.G. et al. (2017). In vivo three-photon imaging of activity of GcamP6-labeled neurons deep in intact mouse brain. *Nature Methods*, 14(4), 388–390. doi: 10.1038/nmeth.4183.

Paiè, P. et al. (2016). Selective plane illumination microscopy on a chip. *Lab Chip*, 16, 1556–1560. doi: 10.1039/C6LC00084C.

Papagiakoumou, E. et al. (2020). Scanless two-photon excitation with temporal focusing. *Nature Methods*, 17(June), 571–581. doi: 10.1038/s41592-020-0795-y.

Planchon, T.A. et al. (2011). Rapid three-dimensional isotropic imaging of living cells using Bessel beam plane illumination. *Nature Methods*, 8(5), 417–423. doi: 10.1038/NMETH.1586.

Ponjavic, A. et al. (2018). Single-molecule light-sheet imaging of suspended T cells. *Biophysical Journal*, 114(9), 2200–2211. doi: 10.1016/j.bpj.2018.02.044.

Preibisch, S. et al. (2010). Software for bead-based registration of selective plane illumination microscopy data. *Nature Methods*, 7(6), 418–419. doi: 10.1038/nmeth0610-418.

Prevedel, R. et al. (2014). Simultaneous whole-animal 3D imaging of neuronal activity using light-field microscopy. *Nature Methods*, 11(7). doi: 10.1038/nmeth.2964.

Prevedel, R. et al. (2016). Fast volumetric calcium imaging across multiple cortical layers using sculpted light. *Nature Methods*, 13(12), 1021–1028. doi: 10.1038/nmeth.4040.

Quirin, S. et al. (2014). Simultaneous imaging of neural activity in three dimensions. *Frontiers in Neural Circuits*, 8(April), 29. doi: 10.3389/fncir.2014.00029.

Roth, S. et al. (2013). Optical photon reassignment microscopy (OPRA). *Optical Nanoscopy*, 2(5). doi: 10.1186/2192-2853-2-5.

Schrödel, T. et al. (2013). Brain-wide 3D imaging of neuronal activity in *Caenorhabditis elegans* with sculpted light. *Nature Methods*, 10(10). doi: 10.1038/nmeth.2637.

Sheppard, C.J.R. (1988). Super-resolution in confocal imaging. *Optik (Jena)*, 80(2), 53–54. doi: 10.1117/12.967201.

Siedentopf, H. and Zsigmondy, R. (1902). Uber Sichtbarmachung und Größenbestimmung ultramikoskopischer Teilchen, mit besonderer Anwendung auf Goldrubingläser. *Annalen der Physik*, 11, 433.

Swoger, J. et al. (2007). Multi-view image fusion improves resolution in three-dimensional microscopy. *Optics Express*, 15(13), 8029. doi: 10.1364/oe.15.008029.

Ta, D.-M. et al. (2023). Label-free image scanning microscopy for kHz super-resolution imaging and single particle tracking. *Optics Express*, 31(22), 36420. doi: 10.1364/oe.504581.

Ueda, H.R. et al. (2020). Whole-brain profiling of cells and circuits in mammals by tissue clearing and light-sheet microscopy. *Neuron*, 106(3), 369–387. doi: 10.1016/j.neuron.2020.03.004.

Vettenburg, T. et al. (2014). Light-sheet microscopy using an Airy beam. *Nature Methods*, 11(4), 541–544. doi: 10.1038/nmeth.2922.

Voigt, F.F. et al. (2019). The mesoSPIM initiative: Open-source light-sheet microscopes for imaging cleared tissue. *Nature Methods*, 16(November). doi: 10.1038/s41592-019-0554-0.

Voleti, V. et al. (2019). Real-time volumetric microscopy of in vivo dynamics and large-scale samples with SCAPE 2.0. *Nature Methods*, 16(10), 1054–1062. doi: 10.1038/s41592-019-0579-4.

Wan, Y. et al. (2019). Light-sheet microscopy and its potential for understanding developmental processes. *Annual Review of Cell and Developmental Biology*, 35, 655–681. doi: 10.1146/annurev-cellbio-100818-125311.

Weigelin, B. et al. (2016). Third harmonic generation microscopy of cells and tissue organization. *Journal of Cell Science*, 129(2), 245–255. doi: 10.1242/jcs.152272.

Wilson, T. (2011). Resolution and optical sectioning in the confocal microscope. *Journal of Microscopy*, 244(2), 113–121. doi: 10.1111/j.1365-2818.2011.03549.x.

Wilson, T. and Sheppard, C.J.R. (1984). *Theory and Practice of Scanning Optical Microscopy*. Academic Press, London. doi: 10.1364/josaa.4.000551.

Wu, Y. et al. (2011). Inverted selective plane illumination microscopy (iSPIM) enables coupled cell identity lineaging and neurodevelopmental imaging in Caenorhabditis elegans. *Proceedings of the National Academy of Sciences of the USA*, 108(43), 17708–17713. doi: 10.1073/pnas.1108494108.

Yang, B. et al. (2019). Epi-illumination SPIM for volumetric imaging with high spatial-temporal resolution. *Nature Methods*, 16(6), 501–504. doi: 10.1038/s41592-019-0401-3.

Yaroslavsky, A.N. et al. (2002). Optical properties of selected native and coagulated human brain tissues. *Physics Medicine & Biology*, 47, 2059–2073. doi: 10.1088/0031-9155/47/12/305.

York, A.G. et al. (2012). Resolution doubling in live, multicellular organisms via multifocal structured illumination microscopy. *Nature Methods*, 9(7), 749–754. doi: 10.1038/nmeth.2025.

York, A.G. et al. (2013). Instant super-resolution imaging in live cells and embryos via analog image processing. *Nature Methods*, 10(10), 1–9. doi: 10.1038/nmeth.2687.

Zhao, Z.W. et al. (2014). Spatial organization of RNA polymerase II inside a mammalian cell nucleus revealed by reflected light-sheet superresolution microscopy. *Proceedings of the National Academy of Sciences of the USA*, 111(2), 681–686. doi: 10.1073/pnas.1318496111.

7

Structured Illumination Microscopy

Alexandra FRAGOLA

Institut des Sciences Moléculaires d'Orsay, Département de Physique, UMR 8214, CNRS, Université Paris Saclay, France

7.1. Introduction

Similarly to STED microscopy and single-molecule localization microscopies, structured illumination microscopy (or SIM) is a fluorescence imaging technique that allows the diffraction limit to be exceeded and thus improves the spatial resolution in all three directions compared to the conventional microscope. In its linear version, SIM provides a gain of about two and a resolution of about a hundred nanometers can be achieved in practice. In its nonlinear version, due to the saturation of the fluorescence excitation, the resolution of SIM is about a few tens of nanometers and is comparable to other super-resolution microscopy techniques. Even if in its linear version, SIM microscopy only provides a modest gain of about two in spatial resolution. The low power density required for excitation and the time resolution of about a fraction of a second means that it can be regarded as a technique of choice for high-resolution 3D imaging of dynamic phenomena in living cells.

This chapter explains the principle of SIM, its experimental implementation, the reconstruction of the SIM image and presents a few application examples in biological imaging. Further details can be found in the many scientific papers, many of which are cited in Heintzmann and Huser (2017) and describe experimental developments of SIM, reconstruction algorithms, applications or recent developments, some of which are mentioned in this book.

Photonic Imaging for Biology,
coordinated by Jean-Baptiste SIBARITA.

7.2. The principle of structured illumination microscopy

7.2.1. *Prerequisites: the Moiré effect*

The Moiré effect is a contrasting effect resulting from the superimposition of two patterns of light and dark lines. Consider, for example, two sine patterns with similar spatial frequencies and different orientations, as shown in Figure 7.1. The superposition of these two patterns reveals a contrast modulation at a lower frequency, related to the difference of the two spatial frequencies of the two patterns and in a direction that depends on their relative orientations (see Figure 7.1A and 7.1B). This phenomenon can be observed on a daily basis, as shown in Figure 7.1C.

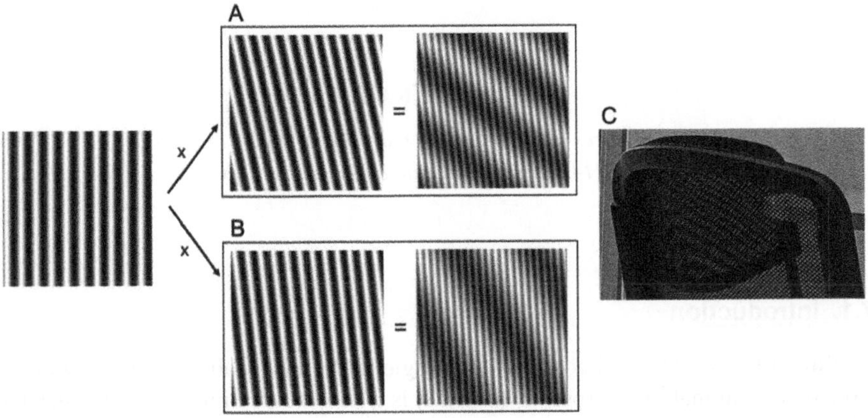

Figure 7.1. *Principle and examples of the Moiré effect. The left sine pattern is superimposed on a similar frequency sine pattern inclined at two different angles (images on the left side of insets A and B). The resulting images show a low-frequency modulation whose orientation and frequency depend on those of the initial patterns (images on the right side of insets A and B). Image C shows an example of the Moiré effect observed in everyday life, here due to the superimposition of chair backs with checkered fabric. For a color version of this figure, see www.iste.co.uk/ sibarita/photonic.zip*

7.2.2. *Structured illumination to exceed the diffraction limit in fluorescence microscopy*

Consider as an object or sample, a variable frequency pattern (see Figure 7.2A(top)), whose spatial frequency spectrum is represented in Figure 7.2A(bottom). It can be distinguished that:

– At the center of the spectrum, the low spatial frequencies that correspond in the sample to the slow contrast variations of the pattern with the space dimension, namely the upper-left part of the pattern.

– Away from the center of the spectrum, the high spatial frequencies, which correspond in the sample to the fast variations in the contrast of the pattern with the space dimension, namely the lower-right part of the pattern. More generally, the high spatial frequencies of the spectrum of a sample correspond to the *details* of the object observed by microscopy.

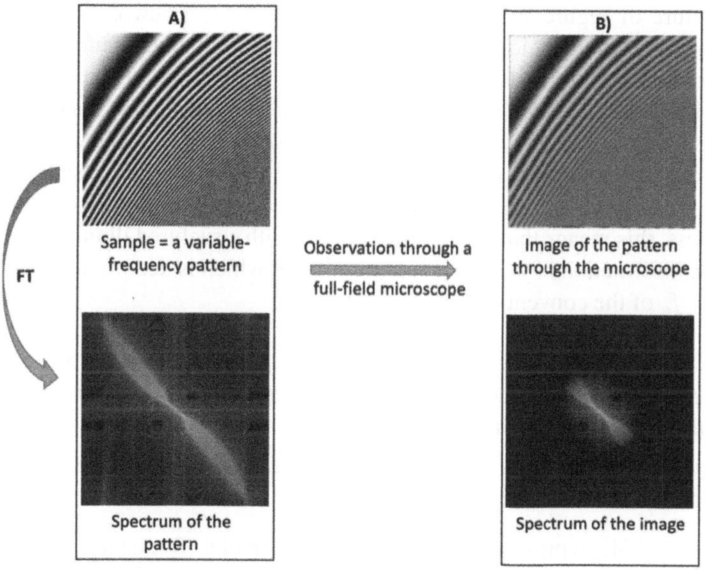

Figure 7.2. *Impact of the lateral spatial resolution limit on the image of a sample through a conventional microscope. A) (top) The test sample is a sine pattern of variable frequency; (bottom) 2D spectrum of the spatial frequencies of the sample. B) (top) Image of the test sample observed through a conventional microscope; (bottom) 2D spectrum of the spatial frequencies of this image, which highlights the loss of high frequencies beyond the disk of radius equal to the cut-off frequency of the microscope*

However, the spatial resolution of an optical microscope, fixed among other things by the numerical aperture of the objective, corresponds in Fourier space to a cut-off frequency, denoted by f_c (see Chapter 1 for more details). Through a conventional optical microscope in which the illumination is homogeneous, the sample details (here the variable pattern) smaller than the resolution cannot be observed (see Figure 7.2B(top)). In Fourier space, this corresponds to the fact that the spatial frequencies of the sample greater than f_c, i.e. located outside the support of the optical transfer function of the microscope (a disk of radius equal to f_c), are lost (see Figure 7.2B(bottom)).

How should the Moiré effect be used to improve the spatial resolution of an optical microscope?

We now illuminate the variable test pattern with structured illumination, for example, a sine pattern with a frequency below f_c. This amounts to superimposing the two patterns, thus multiplying the variable pattern at each point (the sample) (Figure 7.3A) by the sine pattern (illumination) (Figure 7.3B): the result is shown in the picture of Figure 7.3C. Due to the Moiré effect, a new contrast modulation appears at lower frequencies, indicated by the red arrow on the picture in Figure 7.3C and corresponding to the difference between the frequencies of the variable pattern and the frequency of the structured illumination. Compare this image to the one obtained with a conventional microscope (see Figure 7.3C vs. Figure 7.2B): where with homogeneous illumination, the modulation was no longer visible, a contrast can now be observed, even if it does not exactly correspond to that of the object. Nonetheless, this means that the image obtained with structured illumination contains additional information beyond the resolution limit, which is determined by the cut-off frequency f_c of the conventional microscope.

To better understand the phenomenon, let us observe the spectra of the sample *variable pattern* and the structured illumination, as well as that of the image obtained through the microscope in Figure 7.3:

– Figure 7.3D represents the spectrum of spatial frequencies of the variable pattern (the sample).

– Figure 7.3E represents the spectrum of spatial frequencies of sinusoidal illumination, namely three peaks at $f = +f_{modulation}$ and $f = -f_{modulation}$, as well as at $f = 0$ because the average value of the image is not zero.

– Figure 7.3F represents the spectrum of the image of the variable pattern illuminated by sinusoidal illumination through the microscope: the multiplication of the object by a structured illumination in real space corresponds to the convolution product of their spectra in Fourier space. This thus amounts to "copying" the spectrum of the object at the frequencies $f = 0$, $f_{modulation}$ and $-f_{modulation}$.

Finally, we compare the images of the variable pattern through the microscope, as well as their spectra, when the object is illuminated with homogeneous illumination (see Figure 7.2B) or with structured illumination with a sinusoidal pattern (see Figure 7.3C and 7.3F). Figure 7.3C and 7.3F clearly show (indicated by red arrows) that the picture obtained with structured illumination provides access to information about high spatial frequencies beyond the cut-off frequency of the microscope, nonexisting when the sample is illuminated with homogeneous illumination.

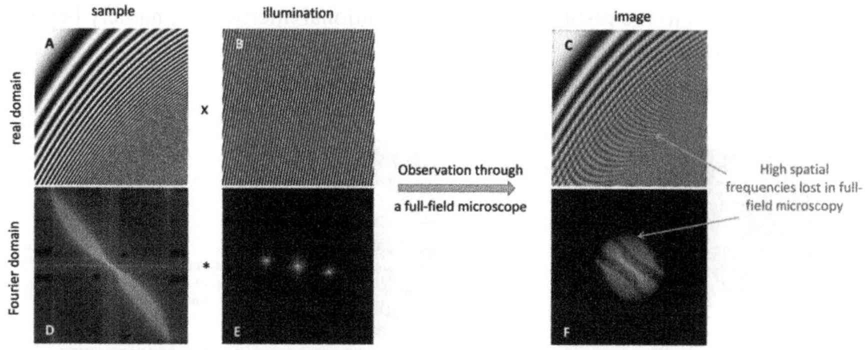

Figure 7.3. *Principle of structured illumination microscopy. Sample A is a variable frequency sine pattern illuminated by a sine pattern; B is a pattern with a high frequency that is close to the cut-off frequency of the conventional microscope. Superimposition C shows a low-frequency modulation, indicated by the red arrow, where there was no modulation in image B (top) of Figure 7.2. For a color version of this figure, see www.iste.co.uk/sibarita/photonic.zip*

Therefore, by illuminating the sample with a sinusoidal pattern, its image through the microscope contains information beyond the resolution obtained with homogeneous illumination, due to the Moiré effect between the illumination frequency and the high frequencies of the sample. To obtain a super-resolved image of the object, it is then necessary to "reposition" the high frequencies indicated by the red arrows in Figure 7.3F at the right place in the spectrum, which is tantamount to *demodulating* the picture.

7.3. Reconstructing the 2D SIM image

The reconstruction of the SIM image proposed by Gustafsson (2000) and Heintzmann (1998) consists of isolating the different parts of the spectrum of the image recorded by the camera in order to "put them back in the right place". To this end, for a given orientation of the pattern at illumination, it has to be translated in its direction of modulation to three positions shifted by a third of a period (or out of phase in Fourier space) and to acquire the three corresponding images with the camera. A proper combination of these images, described in Figure 7.4, then results in the separation of the spectrum parts: the one that corresponds to the spectrum obtained by conventional full-field microscopy, already correctly placed around the frequency $f = 0$, and the two pieces beyond the cut-off frequency f_c, to be placed in the right place (which corresponds to a demodulation in direct space). As shown in Figure 7.4 in step 3, the transfer function obtained after this treatment is then extended in the direction of structuring the illumination by a factor related to its frequency (see section 7.4.2). The spatial resolution is therefore improved in this direction compared to the resolution of the conventional microscope, namely beyond the diffraction limit.

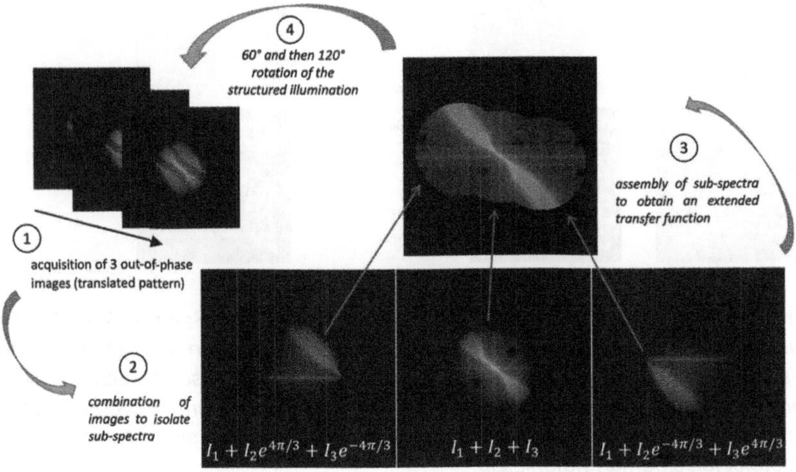

Figure 7.4. *Reconstruction principle of a SIM image. This figure shows, in Fourier space, the main steps during the reconstruction of a SIM image. This approach is one that was originally proposed and remains the most commonly used in SIM microscopy. For a color version of this figure, see www.iste.co.uk/sibarita/photonic.zip*

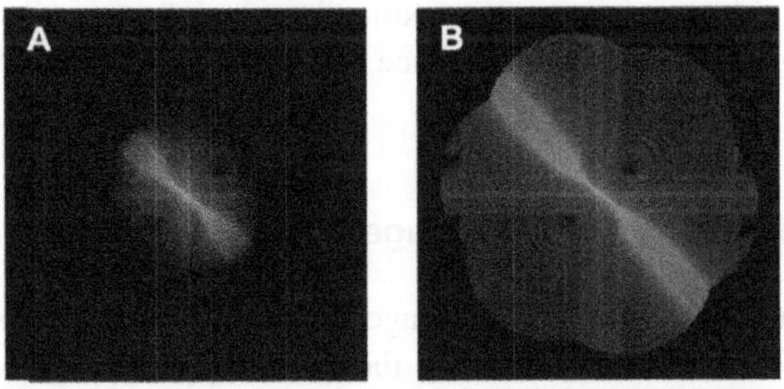

Figure 7.5. *Comparison of the transfer function of the conventional microscope (A) to that of the SIM microscope (B)*

To obtain an extension of the transfer function in all directions, it is necessary to rotate the illumination pattern and to repeat the previous operation (i.e. for each orientation, three positions or phases of the illumination pattern = three images acquired). In practice, three pattern orientations at 60° from one another were used to obtain a gain in quasi-isotropic resolution, as shown by the spectrum of the reconstructed SIM image in Figure 7.5, which also allows us to compare the transfer functions of a conventional full-field microscope and a SIM microscope. To obtain a SIM image, nine images have therefore to be acquired (three images in three pattern orientations). By taking into account the redundancy of information in these nine images, the number of images to be acquired can be reduced to a minimum of seven.

7.4. Experimental implementation of the 2D SIM

7.4.1. *The setup: how to illuminate the sample with a sine pattern?*

To transform a conventional full-field fluorescence microscope into a structured illumination fluorescence microscope, only the excitation pathway needs to be modified in order to allow structured illumination of the sample (for instance, in this case with a sinusoidal pattern). The detection pathway remains unchanged though. The excitation pattern is obtained by making two coherent laser beams interfere directly in the sample, resulting in achieving a sinusoidal modulation with a contrast equal to or very close to 1. The two beams are generally obtained by diffraction of the laser source by a pattern, keeping only the $+1$st and -1st orders. This pattern can be a physical pattern, i.e. an element consisting of alternating transparent and opaque bands, placed in a conjugate plane of the focus plane. As seen earlier, to reconstruct the SIM image and obtain an isotropic resolution, this pattern ought to be laterally shifted and rotated. The latter is therefore placed on a mechanical support that translates and rotates, which will limit the speed of image acquisition between the different orientations of the pattern. A more common alternative solution consists of using a spatial light modulator (often referred to as SLM) to create the pattern. The use of an SLM has several advantages: the absence of mechanical movement and the ability to easily change the spatial frequency of the pattern, its position and its orientation, at rates of up to several hundred Hertz. It is indeed convenient to be able to easily modify the frequency of the structured illumination according to the sample, in particular to decrease it when the image on the camera presents a low signal-to-noise ratio, even if this is at the expense of the gain in spatial resolution. On the other hand, using an SLM imposes a direction of linear polarization of the laser beam used. It should also be noted that by reducing the spatial coherence of the laser, for example, by means of a rotating diffuser, it is possible to limit the formation of interference fringes to the depth of field of the objective lens, and thus to structure the illumination only in the focus plane.

7.4.2. *A resolution gain limited to a factor of 2*

Regardless of the implementation chosen to structure the illumination, the spatial frequency generated in the sample will be restricted by the resolution of the microscope objective and therefore its numerical aperture, and the excitation and emission wavelengths of the fluorescent labeling of the sample. Therefore, the highest possible illumination frequency is equal to the cut-off frequency of the objective lens:

$f_{illum-max} = f_c$. This is why the highest spatial frequency of the sample $f_{ech-max}$ that can be transmitted by the structured illumination microscope is such that:

$$f_{ech-max} - f_{illum-max} = f_c$$

$$\text{that is } f_{ech-max} = 2f_c$$

The lateral spatial resolution in SIM that can be achieved is therefore at best twice that of a conventional microscope with homogeneous illumination: the resolution gain of the SIM is limited to a factor of 2.

In practice, the resolution gain rarely reaches a factor of 2, but is close to it, because the illumination frequency is often lower than the cut-off frequency f_c of the objective lens. Indeed, the reconstruction of SIM images requires an estimation of the parameters of the illumination pattern with sub-pixelic accuracy, thereby a sufficient signal-to-noise ratio and fringe contrast. One way to improve them consists of decreasing the spatial frequency of the excitation pattern, even if this leads to a decrease in resolution gain. Given that the fluorescence emission is incoherent, the transfer function of the microscope during emission is the incoherent transfer function described in Chapter 1. The contrast of a periodic object, such as here the pattern of sinusoidal excitation, decreases as the illumination frequency approaches the cut-off frequency f_c.

7.4.3. *Optical sectioning for 2D SIM: application to biological imaging*

Structured illumination microscopy as we have just described, referred to as 2D SIM, therefore makes it possible to improve the lateral spatial resolution, namely in the focus plane, at most by a factor of 2. On the other hand, the axial resolution is always equal to the depth of field of the objective lens, as for a conventional microscope (see Chapter 1). Without optical sectioning, for a thick sample, the fluorescence from the planes outside the focus plane can therefore induce a strong background signal, which will degrade both the contrast and the signal-to-noise ratio of the image on the camera, and as such degrade the estimation of the pattern parameters during the reconstruction of the SIM images. Moreover, if the reconstruction algorithm does not take into account the presence of an "out-of-focus" background signal, unmodulated by the illumination pattern, this can generate artifacts such as the presence of residual modulation in the reconstructed image.

To overcome this problem, 2D SIM was combined with full-field microscopy techniques enabling optical sectioning (Vermeulen et al. 2015), such as total internal reflection fluorescence microscopy (TIRF) (Kner et al. 2009). In this case, the spatial resolution can be isotropic, of about a hundred nanometers, but is restricted to imaging structures located near the surface of the glass slide. The combination of SIM with optical sectioning has led, for example, to the observation of tubulin filaments in living cells over time (see for example figure 1 in Kner et al. 2009) or to the visualization of neurons in *C. elegans* worms, also living (Vermeulen et al. 2015).

Alternative approaches for reconstructing SIM images, such as the one based on solving the inverse problem (Orieux and Fragola 2017), make it possible to take into account the "background signal" component from the outset and to subtract it from the raw images. This original algorithm is successfully able to obtain 2D SIM images of mitochondrial ridges in living cells and observing their dynamics when they are subjected to photo-induced stress, as shown in Figure 7.6.

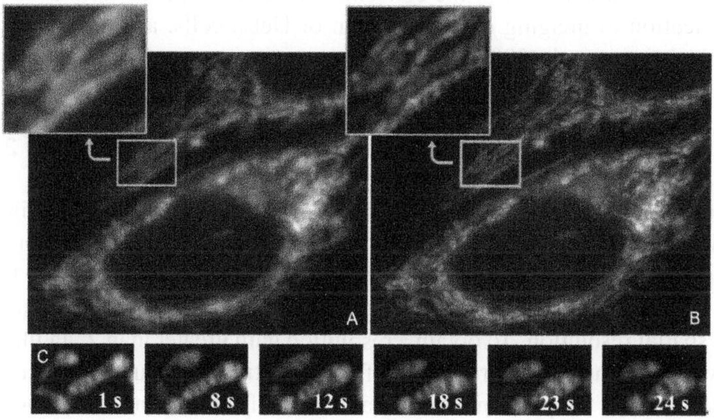

Figure 7.6. *Dynamic SIM imaging of mitochondria in living HeLa cells under photoinduced stress. Comparison of the conventional image (A) with the SIM image (B) and enlargements in the insets showing the gain in spatial resolution. (C) SIM images over time of the same mitochondria: influence of stress on their structure. Results obtained by V. Loriette, A Fragola (LPEM, ESPCI) and S. Bonneau (LJP, SU). For a color version of this figure, see www.iste.co.uk/sibarita/photonic.zip*

7.5. 3D SIM

To improve the spatial resolution beyond the diffraction limit in the three spatial directions and thus obtain a 3D SIM microscope, it is necessary to structure the illumination along the axial direction, denoted by z, in addition to the structuring in the focus plane, denoted by xy. This can be easily achieved by keeping the three laser beams diffracted by the SLM of +1, 0 and -1 orders. When these interfere in the sample, the resulting structuring corresponds to sinusoidal fringes in the xy planes as before, as well as in the z direction. For a better understanding of this, refer to figure 3 in Gustafsson et al. (2008), which shows a XZ view of the structured illumination in the sample).

Similar to 2D SIM, multiple images have to be acquired at each depth (or z-position) in the sample. The 3D image stack is usually achieved by moving the sample along the optical axis. The reconstruction of a 3D SIM image nevertheless requires the acquisition of an increased number of images compared to the 2D SIM: 15 images per plane, which correspond to five lateral positions of the pattern (or phases when reasoning in Fourier space) in three orientations at 60° from each other. A reconstruction algorithm specific to 3D SIM allows us to obtain a volume of SIM images with a gain in spatial resolution with an isotropic factor of 2, in the three spatial directions (Shao et al. 2011). The transfer function of the 3D SIM microscope compared to that of the conventional microscope, as well as an example of an application to imaging of cytoskeleton of HeLa cells, are shown in figures 1 and 6 of Gustafsson et al. (2008).

7.6. Saturated SIM, or SSIM: a resolution gain above a factor of 2

The resolution of SIM is related to the ability to "bring back" the high spatial frequencies of the sample into the support of the transfer function of the conventional microscope, taking advantage of the Moiré effect between them and the frequency of the illumination modulation. Therefore, the higher the illumination frequency, the better the spatial resolution. To obtain a resolution gain greater than 2, an illumination frequency beyond the cut-off frequency of the objective lens must be generated. This is possible by using the phenomenon of saturation of fluorescence excitation. Due to this nonlinear phenomenon, the excitation pattern, which is purely sinusoidal and has a frequency of f_{illum}, becomes a saturated sinusoid at emission, which is closer to a square signal when the fluorescence is increasingly saturated. Its spectrum therefore consists of the frequency f_{illum} and its harmonics, as shown in Figure 7.7a and 7.7b (Gustafsson 2005). It should be noted that the number of raw images to be acquired then significantly increases, in order to be able to "fill" in an almost isotropic way the support surface of the extended transfer function, the diameter of which increases in proportion to the number of harmonics used during the reconstruction. In the example shown in Figure 7.7c, 7.7d and 7.7e, where saturation enables access to three harmonics, it is necessary to acquire nine images of the laterally shifted pattern (nine phases) in 12 orientations, i.e. 108 images in total.

SSIM was published by Gustafsson (2005) and demonstrated with a 2D test sample composed of fluorescent beads with a 50-nm diameter (see Figure 7.7f and 7.7g), with a much better resolution gain than in linear SIM. However, saturating fluorescence requires a high light power density at excitation, which is even higher if we are willing to generate a large number of harmonics in the sample, which can be incompatible with the imaging of living biological samples and can induce strong photobleaching of fluorophores, which in turn limits the number of images that can be acquired. In a paper published in 2015, based on the use of photo-activatable fluorescent molecules, Betzig and his team succeeded in obtaining SSIM images in living cells with resolutions of about 40 nanometers (Li et al. 2015).

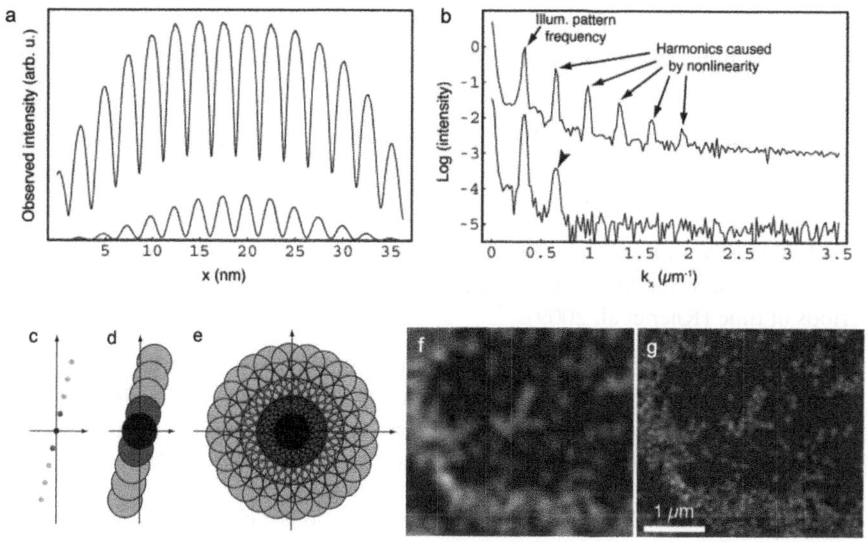

Figure 7.7. *Saturated SIM, illustrated with figures from reference (Gustafsson 2005; Copyright (2005) National Academy of Sciences, USA). (a) Image profiles of a thin fluorescent layer illuminated by a modulated Gaussian laser beam with two different energy densities that saturate the fluorescence (density of the bottom curve < density of the top curve). (b) Fourier transforms corresponding to the profiles shown in (a), showing one harmonic for low-energy density and five harmonics for high-energy density (arrows). The curves have been vertically separated for clarity. (c) Spatial frequencies of the emission pattern for a modulation direction: the frequency at the origin represents the average value, the two peaks close to the origin are the frequency of the illumination pattern (namely at excitation) and the other six peaks correspond to the three harmonics resulting from fluorescence saturation. (d) Extension of the transfer function after reconstruction for a single modulation direction. (e) Transfer function obtained after reconstruction with all harmonics and orientations of the pattern in saturated SIM with three harmonics. (f–g) Comparison of linear SIM images and saturated SIM images of 50-nm fluorescent beads, demonstrating the gain in lateral spatial resolution*

7.7. Conclusion and perspective

In its linear version, SIM was initially not as successful as other super-resolution microscopy techniques such as STED or SMLM microscopies, partly because the gain in resolution is not as high, being limited to a factor of 2, and partly because the reconstruction of SIM images is complex. However, SIM microscopy experienced a strong renewed interest about 15 years ago, because it has definite advantages over other super-resolved microscopy techniques.

– No specific labeling required with dedicated fluorophores, and sometimes delicate application protocols: SIM allows for the observation of fluorescent biological samples already studied within a team without developing new preparation protocols.

– The SIM setup can easily be used with a multi-wavelength excitation module. It is therefore compatible with most fluorescent labeling and multi-color imaging techniques (Fiolka et al. 2012).

– Linear SIM is a full-field microscopy technique, so the light power density at excitation remains moderate and compatible with *in vivo* dynamic imaging over long periods of time (Kner et al. 2009).

– Finally, a SIM setup is based on the implementation of a structured illumination module on a full-field microscope, thus only on the excitation pathway. It is therefore easy to couple it with other modalities such as multifocus imaging for fast 3D SIM (Abrahamsson et al. 2017) or supercritical fluorescence microscopy *SAF*.

This is why linear SIM microscopy, even with a modest resolution gain of a factor of 2, is now widely used by biologists, mainly for cellular imaging as well as to observe small, relatively transparent samples such as *C. elegans* worms. This success is linked to its isotropic spatial resolution of about a hundred nanometers, combined with a sub-second time resolution, compatible with super-resolved dynamic 3D imaging of living samples, as well as to the development of easy-to-use commercial systems and the availability of free reconstruction algorithms such as the *FairSIM* project (Müller et al. 2016).

SIM microscopy is also very successful in research, to miniaturize it (Helle et al. 2020) or to combine it with other microscopy modalities, such as adaptive optics (Thomas et al. 2015) or light-sheet microscopy (Chang et al. 2017; Chen et al. 2023), and thus improve its performance. There is also an alternative approach where the illumination is no longer structured by a sinusoidal pattern but with a random speckle pattern. This approach, called RIM for *random illumination microscopy* (Mangeat et al. 2021), has the advantage of not needing the sensitive step of estimating the pattern parameters in the acquired data, since the statistical properties of the speckle pattern are known and sufficient to reconstruct the super-resolved image. This facilitates the reconstruction of the images, although the trade-off resides in the large number of raw images to be acquired (several hundred images), which limits the time resolution of RIM for dynamic imaging of living samples. Nevertheless, RIM has a number of advantages: the generation of the speckle pattern and its movement in the sample to acquire the different images is simple to implement. There is a free reconstruction algorithm, called *AlgoRIM*, and the speckle pattern appears to be more resistant to the scattering of biological tissue, paving the way for SIM microscopy deep in the small animal.

7.8. References

Abrahamsson, S., Blom, H., Agostinho, A., Jans, D.C., Jost, A., Müller, M., Nilsson, L., Bernhem, K., Lambert, T.J., Heintzmann, R. et al. (2017). Multifocus structured illumination microscopy for fast volumetric super-resolution imaging. *Biomedical Optics Express*, 8(9), 4135.

Chang, B.-J., Perez Meza, V.D., Stelzer, E.H.K. (2017). csiLSFM combines light-sheet fluorescence microscopy and coherent structured illumination for a lateral resolution below 100 nm. *Proceedings of the National Academy of Sciences*, 114(19), 4869–4874.

Chen, B., Chang, B.-J., Roudot, P., Zhou, F., Sapoznik, E., Marlar-Pavey, M., Hayes, J.B., Brown, P.T., Zeng, C.-W., Lambert, T. et al. (2023). Resolution doubling in light-sheet microscopy via oblique plane structured illumination. *Nature Methods*, 19(11), 1419–1426.

Fiolka, R., Shao, L., Rego, E.H., Davidson, M.W., Gustafsson, M.G.L. (2012). Time-lapse two-color 3D imaging of live cells with doubled resolution using structured illumination. *Proceedings of the National Academy of Sciences*, 109(14), 5311–5315.

Gustafsson, M.G.L. (2000). Surpassing the lateral resolution limit by a factor of two using structured illumination microscopy. *Journal of Microscopy*, 198(Part 2), 82–7.

Gustafsson, M.G.L. (2005). Nonlinear structured-illumination microscopy: Wide-field fluorescence imaging with theoretically unlimited resolution. *Proceedings of the National Academy of Sciences*, 102(37), 13081–13086.

Gustafsson, M.G.L., Shao, L., Carlton, P.M., Wang, C.J.R., Golubovskaya, I.N., Cande, W.Z., Agard, D.A., Sedat, J.W. (2008). Three-dimensional resolution doubling in wide-field fluorescence microscopy by structured illumination. *Biophysical Journal*, 94(12), 4957–4970.

Heintzmann, R. and Cremer, C. (1998). Laterally modulated excitation microscopy: Improvement of resolution by using a diffraction grating. *Proceedings of the SPIE*, 3568, 185–196.

Heintzmann, R. and Huser, T. (2017). Super-resolution structured illumination microscopy. *Chemical Reviews*, 117(23), 13890–13908.

Helle, Ø.I., Dullo, F.T., Lahrberg, M., Tinguely, J.-C., Hellesø, O.G., Ahluwalia, B.S. (2020). Structured illumination microscopy using a photonic chip. *Nature Photonics*, 14(7), 431–438.

Kner, P., Chhun, B.B., Griffis, E.R., Winoto, L., Gustafsson, M.G.L. (2009). Super-resolution video microscopy of live cells by structured illumination. *Nature Methods*, 6(5), 339–342.

Li, D., Shao, L., Chen, B.-C., Zhang, X., Zhang, M., Moses, B., Milkie, D.E., Beach, J.R., Hammer, J.A., Pasham, M. et al. (2015). Extended-resolution structured illumination imaging of endocytic and cytoskeletal dynamics. *Science*, 349(6251), aab3500.

Mangeat, T., Labouesse, S., Allain, M., Negash, A., Martin, E., Guénolé, A., Poincloux, R., Estibal, C., Bouissou, A., Cantaloube, S. et al. (2021). Super-resolved live-cell imaging using random illumination microscopy. *Cell Reports Methods*, 1(1), 100009.

Müller, M., Mönkemöller, V., Hennig, S., Hübner, W., Huser, T. (2016). Open-source image reconstruction of super-resolution structured illumination microscopy data in ImageJ. *Nature Communications*, 7(1), 10980.

Orieux, F., Loriette, V., Olivo-Marin, J.-C., Sepulveda, E., Fragola, A. (2017). Fast myopic 2D-SIM super resolution microscopy with joint modulation pattern estimation. *Inverse Problems*, 33(12) 125005. doi: 10.1088/1361-6420/aa8f5b.

Shao, L., Kner, P., Rego, E.H., Gustafsson, M.G.L. (2011). Super-resolution 3D microscopy of live whole cells using structured illumination. *Nature Methods*, 8(12), 1044–1046.

Thomas, B., Wolstenholme, A., Chaudhari, S.N., Kipreos, E.T., Kner, P. (2015). Enhanced resolution through thick tissue with structured illumination and adaptive optics. *Journal of Biomedical Optics*, 20(2), 026006.

Vermeulen, P., Zhan, H., Orieux, F., Olivo-Marin, J.-C., Lenkei, Z., Loriette, V., Fragola, A. (2015). Out-of-focus background subtraction for fast structured illumination super-resolution microscopy of optically thick samples: Structured illumination microscopy. *Journal of Microscopy*, 259(3), 257–268.

8

STED Imaging of Neuronal Morphology

Veera Venkata Gopala Krishna INAVALLI[1] and Urs Valentin NÄGERL[2]

[1] *Center for Cancer Immunology, University of Southampton, United Kingdom*

[2] *Department of Anatomy and Cell Biology, Universitätsmedizin Göttingen, Georg-August-Universität, Germany*

Progress in microscopy technology has a long history of triggering major advances in neuroscience. Stimulated emission depletion microscopy (STED) is a far-field super-resolution microscopy (SRM) technique based on laser-scanning fluorescence microscopy, which provides high optical sectioning in thick biological samples. STED gives access to the nanoscale organization and dynamics of (sub)cellular structures of neurons and glial cells, which are impossible to resolve using conventional light microscopy. In this chapter, we explain the basic principle of STED microscopy, its instrumentation and several recent advancements that have made it a powerful technique well suited to probe neural morphology in a live setting with nanometric spatial resolution.

8.1. Introduction

Major advancements in neuroscience have often been sparked by developments in microscopy techniques: Golgi staining in the 19th century, electron microscopy, laser-scanning fluorescence microscopy in the 20th century and genetically encoded biosensors and optogenetics in more recent times. Camillo Golgi and Santiago Ramon y Cajal invented new staining methods that allowed them to see individual cells, describe the fine organization of neural tissue and identify subcellular components and organelles; laying the groundwork for modern neuroscience along

Photonic Imaging for Biology,
coordinated by Jean-Baptiste SIBARITA.

the way. The invention of the electron microscope led to unprecedented insights into the ultrastructure of neurons and biological tissue in general.

Light microscopy techniques initially relied on the intrinsic contrast of cellular features or unspecific contrast enhancers, but the invention of dye-conjugated antibodies in the 1940s (Coons and Kaplan 1950) paved the way to imaging cells and subcellular structures with molecular specificity. In the past 30 years, the development and application of genetically encoded fluorescence probes, starting with green fluorescent protein (GFP), as tools for cell labeling and protein tagging in combination with tremendous improvements in light microscopy technology, image processing and analysis have revolutionized biomedical research, in particular neurobiology.

Fluorescence microscopy has become the most widely used imaging technique in neuroscience for exploring the structure and function of brain tissue across a wide range of spatio-temporal scales. This method has many benefits, ranging from relative non-invasiveness, labeling specificity for cell types and subcellular components, high time resolution, to the compatibility with other powerful methodologies, such as electrophysiology and optogenetics (Kubitscheck 2017).

The development of confocal microscopy in the 1980s and two-photon microscopy in the 1990s were milestones that made it possible to obtain high-contrast, narrow depth of focus fluorescence images of live biological tissue, in ex vivo preparations or even inside the intact brain in vivo (Pawley 2006; Sanderson 2019). Through optical sectioning, the confocal microscope enhances image signal-to-noise ratio (SNR), but it suffers when imaging thick samples due to scattering, absorption and out-of-focus fluorescence, which impair image quality and induce photobleaching and phototoxicity. Because of the longer-wavelength excitation, two-photon microscopy mitigates these problems and is capable of imaging deep inside biological tissue. A dditionally, the nonlinearity of two-photon excitation enhances optical sectioning.

Confocal and two-photon microscopy are heavily used in neurobiology to reveal morphological features of cellular structures and their interactions. However, the diffraction of light imposes a hard lower limit on the size of the structures that can be resolved by these techniques. The limit is defined by the famous Abbe resolution formula:

$$\Delta r = \lambda / 2 * NA \quad [8.1]$$

where Δr is the minimal distance between two structures for them to be distinguishable, λ is the wavelength of the light and NA is the numerical aperture of the microscope objective. For instance, visible light with a high NA objective lens has a diffraction limit of approximately 200 nm, which means that biological structures smaller than 200 nm cannot be resolved by regular light microscopy. Many key structural features of neurons and glia cells are not resolvable by confocal or two-photon microscopes. These unresolvable features include filopodia and small dendritic spines, spine necks, postsynaptic densities, synaptic clefts and peri-synaptic astrocytic processes.

The diffraction limit of far-field optics was so engraved in the minds of scientists as an immutable dogma that it stood basically unchallenged for over 100 years. As a positive side effect, perhaps, it forced the development of other powerful microscopy methods to resolve structures beyond the 200-nm limit such as electron microscopy, X-ray crystallography, atomic force microscopy and near-field scanning optical microscopy. Unfortunately, these techniques are incompatible with live-cell imaging and, even for fixed cell imaging, necessitate complex sample preparation. Therefore, the pressure and incentive for developing light-based techniques to study the nanocosm of living matter only grew stronger.

Over the last two decades, conceptual and technical advances have led to the development of "super-resolution microscopy" (SRM) or "nanoscopy" techniques, which made it possible to probe nanoscale structures with photons in the visible range of the light spectrum. By controlling the fluorescent on–off state of the molecules, their spatial resolution can theoretically reach the ultimate limit – the physical size of the fluorescent molecule – and not by pushing the wavelength of light or the numerical aperture of the objective to more extreme values.

SRM can be divided into "deterministic" (also called "coordinate-targeted") versus "stochastic" approaches. The goal of deterministic SRM (STED, RESOLFT) (Hell 2007) is to reduce the size of the fluorescent spot that is scanned across the sample as much as possible. The mechanism to switch off the fluorescence occurs via the process of stimulated emission in the case of STED microscopy and a light-induced change in the fluorophore state in the case of RESOLFT, but it can be any type of process that drives reversible and saturable fluorescence transitions between the "on" and "off" states.

Stochastic super-resolution methods (such as PALM, STORM and uPaint) (Betzig et al. 2006; Rust et al. 2006; Giannone et al. 2010; and for a review: Fürstenberg and Heilemann 2013; Lelek et al. 2021) localize single molecules by imaging them individually and calculating the geometric center of the blurry image spot, which is due to the diffraction of light and corresponds to the point spread function (PSF) of the microscope (see Chapter 1). To make sure only one molecule gives rise to this blurry spot, most fluorophores must be kept in a dark, nonfluorescent "off" state. After individual molecules are switched to the fluorescent "on" state stochastically, they can be localized before photobleaching or transiting back to the dark state. The image acquisition time needs to be long enough to ensure that a sufficient number of the molecules have been switched "on" at least once. The image is reconstructed from the positions calculated for each emitter. The final image resolution depends on the localization precision of each emitter (Δloc) and labeling density:

$$\Delta loc \sim \Delta/\sqrt{n} \qquad [8.2]$$

where n is the number of collected photons, Δ is the full width at half-maximum of the PSF.

SRM approaches have sparked a significant paradigm shift in the conceptualization, design and analysis of experimental procedures. This transformative impact extends not only to practical methodologies but also to the theoretical frameworks applied in tackling complex neurobiological challenges (Masters 2015). In this chapter, we will focus on STED microscopy and explain its basic optical principles and technical implementation as well as several key advancements that have increased its scope and usefulness, making it a viable technique for imaging live brain tissue with nanometric spatial resolution.

8.2. Principle

In traditional laser-scanning microscopy (confocal or two-photon), the excitation beam is focused by the microscope objective to a small focal spot and scanned across the specimen in one, two or three spatial dimensions, typically using movable mirrors to steer the laser beam in x and y and a piezo-electric actuator to lift the objective up and down in z. The emitted fluorescence signal is collected all along the scan by a fast and sensitive light detector and by assigning the emitted signal to the spatial coordinate of the scan (x,y,z) that corresponds to the physical position in the sample, the final image is constructed. The excitation and emission PSFs are diffraction-limited spots (see Figure 8.1(a), top panel), meaning that two fluorescent objects remain indistinguishable if their separation is less than 200 nm. Therefore, the focal spot of a laser-scanning microscope should be as small as possible to get the highest spatial resolution. The fundamental idea of STED microscopy is to shrink the size of the focal excitation spot and thereby improve the spatial resolution. To achieve this shrinkage, STED uses a second laser beam, usually shaped like a doughnut in the lateral (xy plane) and like a chimney in the side view, to de-excite the molecules (turn off) in the outer region of the excitation spot. As a result, fluorescence only emanates from the central part of the doughnut (see Figure 8.1(a), lower panel).

The concept of STED microscopy was first proposed in 1994 (Hell and Wichmann 1994), where the deactivation of fluorophores is achieved through stimulated emission. This process can be understood in terms of the Jablonski two-level energy diagram (see Figure 8.1(b)). Upon absorption of an incoming photon ($h\nu_{ex}$), the molecule is excited from the ground state (S_0) to a higher vibrational level of the first excited state (S_1^{vib}), It then rapidly (within picoseconds) decays to the lowest energy level of S_1 by vibrational relaxation. The molecule typically stays in the S_1 state for a few nanoseconds before it returns to the ground state via spontaneous emission, or it can be actively de-excited through stimulated emission. In spontaneous emission, the molecule goes from the lowest S_1 state to the S_0^{vib} state by releasing a fluorescence photon ($h\nu_{fl}$), which is red-shifted compared to the excitation photon due to the energy loss during vibrational relaxation. During simulated emission, a second incoming photon ($h\nu_{STED}$), which matches the energy gap between the S_1 and S_0^{vib} states, sends the molecule from the excited state to the ground state, reducing the number of molecules in the excited state.

The incident and "harvested" stimulated emission photons have identical properties: their wavelength, polarization, phase and direction of propagation are perfectly matched, leading to their designation as "copy photons". Importantly, the stimulated emission light does not re-excite the same molecule to the excited state S_1 due to the energy mismatch; molecules in the higher vibrational levels of the ground state (S_0^{vib}) rapidly decay to the lowest ground state S_0.

Similar to fluorescence, stimulated emission is a radiative process. However, the wavelength of the stimulated emission laser is at the far-right end of the fluorescence emission spectrum, allowing it to be effectively filtered out with appropriate filters. The laser used to induce stimulated emission is called the STED beam.

The STED beam de-excites the molecules in the outer region of the excitation spot through stimulated emission. A s a result, the fluorescence occurs only from the central region. However, to effectively suppress the fluorescence in the outer region of the doughnut beam, the transition rate induced by stimulated emission must be significantly larger than that of spontaneous emission, $k_{fl} << k_{STED}$. This constraint can be met in two different ways: by increasing the STED laser intensity, which increases the probability of de-excitation of the excited molecules in the outer region; or by increasing the stimulated emission cross section of the dye.

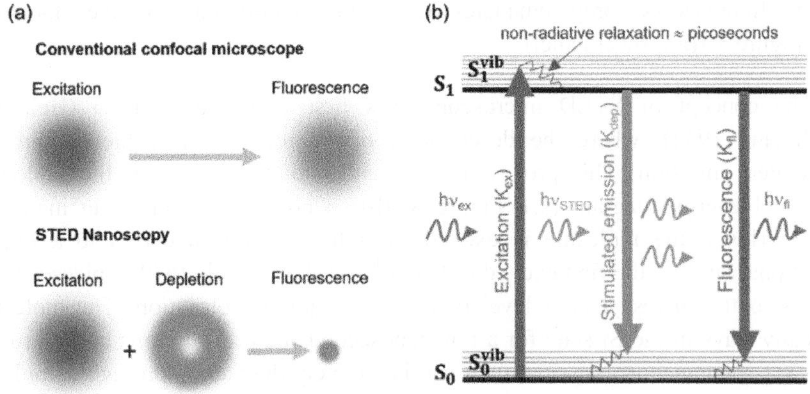

Figure 8.1. *(a) Depiction of a diffraction-limited conventional confocal microscope and super-resolved STED nanoscopy. (b) Jablonski diagram showing the process of excitation, stimulated emission and fluorescence. For a color version of this figure, see www.iste.co.uk/sibarita/photonic.zip*

The STED resolution is derived from a modified version of Abbe's formula [8.1]:

$$\Delta r \approx \frac{\lambda}{2NA\sqrt{1+I_{STED}/I_{Sat}}} \quad [8.3]$$

where λ is the wavelength, NA is the numerical aperture of the objective lens, I_{STED} is the intensity of the STED beam and I_{sat} is the STED intensity at which half of the fluorescence molecules are de-excited. The ratio between the I_{STED}/I_{sat} determines the achievable resolution. For example, if $I_{STED}/I_{sat} = 900$, the theoretical resolution will be improved by about 30 times over the diffraction limit. The saturated intensity of the fluorescent dyes used in STED microscopy can be calculated by the following relation:

$$I_{sat} = h\nu_{STED} / \tau\sigma \quad [8.4]$$

where $h\nu_{STED}$ denotes the STED photon energy (~ 2 eV), σ is the absorption cross section and τ is the fluorescence lifetime ($1/k_{fl}$). For instance, GFP has a lifetime of 2.5 ns and an absorption cross section of approximately 10^{-16}–10^{-17} cm^2. Therefore, the I_{sat} of the GFP is approximately 3 MW/cm^2. I_{STED} values typically range from 10 to 10^3 MW/cm^2 for commonly used fluorophores. The higher the I_{STED}, the higher the achievable spatial resolution, potentially reaching the ultimate limit set by the size of the emitting molecule (Harke et al. 2008).

However, in practical terms, the resolution depends strongly on several other factors as well: the purity of the null (center of the doughnut), which can be affected by optical imperfections (aberrations and incorrect polarization), available STED laser power, the precision of the scanning system and the photophysical and photochemical properties of the fluorophore. The highest resolution reported to date is 6 nm, which was accomplished by imaging nitrogen-vacancy color centers in diamonds (Rittweger et al. 2009).

The two lasers required for STED microscopy can be either two pulsed lasers or two continuous wave (CW) lasers or one pulsed and one CW laser. A supercontinuum laser can also be used to generate both wavelengths, eliminating the need for synchronization between two different laser sources. However, each combination has its pros and cons depending on cost, implementation and applications. For biological applications, pulsed STED is preferable due to its superior SNR and better resolution compared to other combinations with the same STED laser power, and more importantly, long STED pulses (> 150 ps) improve the photostability of the dye compared to short pulses due to the reduce photobleaching. Short laser pulses have multiple interactions with the dye, leading to dark state transition (Dyba and Hell 2003), discussed in the photobleaching section. In principle, stimulated emission can work with any fluorescent dye, but the limitations of available pulsed lasers and the power requirements have restricted the STED implementation to three main wavelengths (STED lasers: 595 nm, 660 nm and 775 nm).

8.3. Instrumentation of STED

After the first experimental demonstration of the STED effect (Klar and Hell 1999), its instrumentation and design have evolved with technological developments (lasers, scanners, detectors, fluorophores, etc.), improving robustness and performance, while reducing complexity and cost. Here, we describe the design of a STED microscope, based on pulsed excitation and pulsed STED lasers, and several technological developments that have made it simpler and more compact for neuro-imaging applications in living cells and tissue. To start, a STED microscope is quite similar in construction and operation to a standard confocal or two-photon laser-scanning fluorescence microscope:

– *Microscope stand*: a home-built STED microscope can be used for live-cell imaging with confocal STED or two-photon STED (see Figure 8.2(a)). It is built around an inverted or upright microscope stand and equipped for electrophysiology to carry out neurobiological experiments in brain slices or living mice. Standard commercial research microscopes can be used for this purpose (Table 8.1). An objective lens with a high numerical aperture should be used to focus the incident light into the sample. The microscope is adapted with a movable sample stage to select the region of interest in the X and Y directions.

– *Lasers*: the most expensive component of a STED microscope is used to be the pulsed STED laser, although as technology advanced, it has become much cheaper. Recent advancements in fiber laser technology, for example, have significantly contributed to this cost reduction. In the early days, bulky Ti:sapphire lasers were used in conjunction with an optical parametric oscillator (OPO) to produce light of the right wavelength and with enough power for stimulated emission. Additionally, the femtosecond pulses from the OPO had to be stretched up to 200–600 ps by glass rods and long polarization-maintaining fibers to facilitate the process of stimulated emission (Blom and Widengren 2017). However, advances in fiber lasers have eliminated these difficulties and made it possible to get suitable pulses directly from the laser source, considerably reducing the cost, maintenance and space needed to implement STED.

We implemented STED in the visible region with a 485-nm diode pulse laser or a two-photon laser for the excitation and a pulsed orange laser for STED de-excitation. Initially, we used a Ti:sapphire laser in combination with an OPO (Chéreau et al. 2015), and later on switched to a picosecond fiber laser (Bancelin et al. 2021). This combination of excitation and de-excitation makes it possible to use fluorescent proteins and dyes such as GFP, YFP, Alexa Fluor 488 and Alexa Fluor 514, which are commonly used to label brain cells.

– *Synchronization unit*: it is crucial to be able to control the relative timing of the pulses from the excitation and depletion lasers, and maximize the effectiveness of stimulated emission to switch off the fluorescence. Usually, it is necessary to delay the STED laser by a couple of hundreds of picoseconds relative to the excitation laser. The STED laser delay can be implemented in two ways: through the optical path or using an electronic trigger. In the optical path method, the STED laser travels more distance than the excitation laser before it reaches the objective, the delay being controlled by moving the mirror on a translational stage. This method requires a complex beam alignment with sophisticated optics. Alternatively, the output signal from one of the lasers (master) is fed to the trigger input of the other laser (slave) via the signal control box (Table 8.1). By controlling the trigger signal timing to the slave, a variable delay between excitation and depletion pulses can be accomplished. Electronic control provides more timing precision and is easy to use on a daily basis (most of the lasers come with master and slave ports and the signal delay boxes are commercially available). We used a custom-built electronic trigger to implement temporal synchronization between excitation and depletion pulses, where the STED laser was the master, and the excitation laser was the slave (see Figure 8.2(a)) (sync unit).

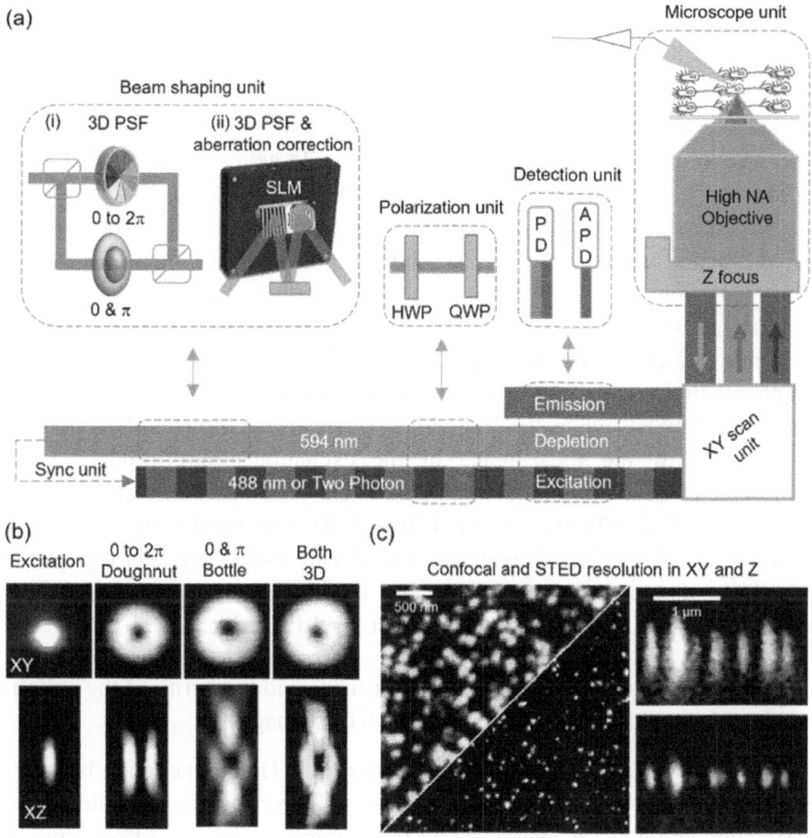

Figure 8.2. *Schematic of a STED microscope setup and point spread functions (PSF) of gold and fluorescent beads. (a) STED setup consists of different parts: laser, synchronization unit, beam shaping, polarization, scan, microscope optics and detection (refer to the text for explanations). (b) PSFs of excitation and depletion beams measured by imaging reflections from a 150-nm gold bead with a photodiode (PD). (c) Confocal and STED resolution measured by imaging 40-nm fluorescent beads, lateral (X, Y) and axial (Z), with an avalanche photodiode (APD). SLM: spatial light modulator; HWP: half-wave plate; QWP: quarter waveplate. For a color version of this figure, see www.iste.co.uk/sibarita/photonic.zip*

– *Shaping of the STED beam*: shaping the depletion beam is a vital part of the STED microscope. The beam shaping can be achieved in two different methods: one using fixed phase plates and the other using dynamic phase devices such as spatial light modulator (SLM) and deformable mirror (DM). In beam shaping with a fixed phase plate, two different phase retardation plates are used for 3D STED implementation, one for lateral resolution (X, Y) and the other for axial (Z) resolution. The beam splitter divides the STED beam into two branches, where each one passes through a different phase plate before being recombined by a second beam splitter (see Figure 8.2(a), beam shaping unit (i)). Due to the destructive interference in the center, the phase plate with a 0–2π helical phase mask produces a doughnut shape beam with a helical charge of 1. The "Top Hat", annular phase mask, with a 0 and π phase delay produces a bottle-shaped beam because of the phase difference between the center and periphery. Beam shaping using SLM involves splitting the device into two parts, with each part loaded with an appropriate phase mask to generate doughnut and bottle-shaped beams in real time (see Figure 8.2(a), beam shaping unit (ii)). The doughnut beam depletes the fluorophores in the same focal plane as the excitation beam, producing the laterally super-resolved PSFs, while the "Top Hat" phase mask depletes the fluorophores above and below the focal plane of the excitation beam, producing axially super-resolved PSFs. The bottle beam also slightly increases the lateral resolution (Wildanger et al. 2009). In our initial 3D STED implementation, we shaped the PSFs using two phase plates (Tønnesen et al. 2018). These were later replaced by an SLM which not only generates PSFs but also helps to correct the aberrations (detailed in the aberration section) (Inavalli et al. 2019; Arizono et al. 2020).

The following factors influence the beam shape and the purity of the null, hence the achievable resolution and the SNR of the STED images:

i) *Spatial alignment*: the spatial alignment of STED and excitation beams in all three dimensions is essential because slight misalignment in any direction degrades the resolution and SNR. A lateral (X, Y) misalignment causes an offset between the excitation peak intensity and the null of the depletion beam in the lateral plane, while an axial (Z) misalignment does the same along the z axis. The spatial alignment is accomplished by looking at back reflections from gold beads in low light settings (100 μW) where the reflected signal can be detected by a photodiode. The PSFs of the excitation and STED beams are measured by scanning the 150-nm gold beads in the XY and XZ directions (see Figure 8.2(b)). Precision optics and the lens position in the excitation path can be used to fine-tune the overlay of both beams.

ii) *Helical charge of the depletion beam*: the STED phase mask converts the plane wavefront into helical wavefront with a charge, and the STED beam can theoretically have any helical charge. For example, a beam with a 0–2π phase has a charge of 1 and a beam with a 0–4π phase has a charge of 2, etc. However, the helical charge has a direct influence on the size of the null and the achievable resolution. A STED beam with a helical charge of 2 requires more laser power to obtain the same resolution than a beam with a helical charge of 1. Most importantly, the helicity and the polarization of the STED beam play a crucial role in the purity of the null and beam shape, discussed in more detail in the section below.

iii) *Depletion beam polarization and helicity*: high NA objectives influence STED beam shape and the purity of the null due to tight focusing effects (concentrating light in a very small area), degrading resolution and SNR as a result (Hao et al. 2010; Brouwer 2016). These effects, however, can be mitigated with the appropriate combination of polarization and helical charge (left, +1 or right, -1). A STED beam with linear polarization has a greater impact on beam shape and null purity. The beam shape elongates in the linear polarization direction, and the null has a residual intensity (see Figure 8.3(a)). On the other hand, circular polarization has no effect on the shape of the beam (see Figure 8.3(b) and (c)), but it requires an appropriate helicity to cancel the residual intensity in the center.

For an absolute null, the helicity of the wavefront should match the handedness of the circular polarization (left or right), which cancels the residual intensity owing to destructive interference (see Figure 8.3(b)) (left circular polarization with left helicity). The opposite directions (right circular polarization with left helicity) lead to constructive interference with strong residual intensity in the center (see Figure 8.3(c)). For instance, a twofold increase in input intensity results in a 70% drop in peak fluorescence, while a fivefold increase causes a 95% loss of signal. Therefore, for maximal resolution and SNR ratios, an appropriate polarization and helicity of the STED beam are essential.

A combination of half ($\lambda/2$) and quarter ($\lambda/4$) waveplates (see Figure 8.2(a)) can be used to tune the input polarization of the STED beam. The $\lambda/2$ plate rotates the plane of polarization, while the $\lambda/4$ plate turns linear polarization into circular polarization. In principle, the $\lambda/4$ plate is enough to produce circular polarization since most pulsed lasers are linearly polarized. However, when combined with a $\lambda/2$ plate, the polarization and handedness can be adjusted more easily.

iv) *Aberrations*: the shape and purity of the null of the 3D-STED beam are affected by optical aberrations such as spherical, coma and astigmatism (Deng et al. 2010). Aberrations are deformations in the wavefront brought on by the imaging system as well as the biological specimen. System aberrations (static) result from imperfections in the optics, faulty alignment and mismatches in the refractive index among the immersion medium, coverslip and imaging medium. When imaging deep within brain tissues, additional aberrations (mostly spherical) induced by the sample also come into play.

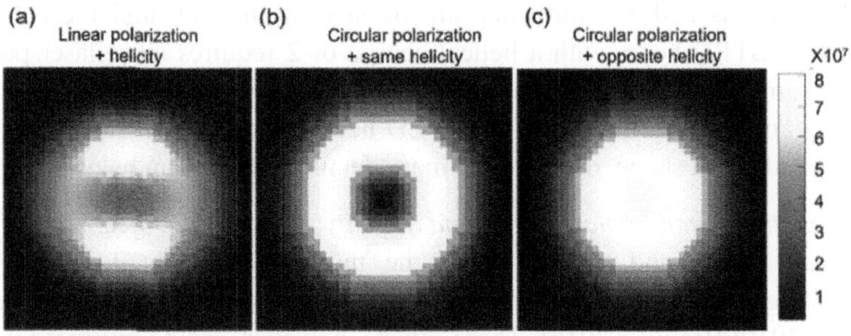

Figure 8.3. *MATLAB simulations of polarization and helicity effect on the depletion beam: (a) linearly polarized doughnut beam does not preserve the beam shape and the null purity; (b) circular polarization with the same helicity preserves the shape and the null; and (c) circular polarization with opposite helicity preserves the shape but very strong residual intensity in the center. Intensity scale bar MW/m^2. For a color version of this figure, see www.iste.co.uk/sibarita/photonic.zip*

Spherical aberrations are the most common problem in imaging systems due to a mismatch of refractive index between the immersion medium and sample. Coma and astigmatism are caused by low-quality optical components and poor alignment of the optics and scanner. However, these aberrations can be minimized by careful design, appropriate immersion medium and cover glass, and by an objective lens equipped with a correction collar. But the sample-induced aberrations are more troublesome and limit the ability to image much beyond the surface of the sample (>10 µm). Unfortunately, sample aberrations cannot be fully corrected using static classical phase plates. Instead, dynamic phase plates can be used for this purpose, such as deformable mirrors or SLMs. The use of SLM is covered in the section on 3D-STED using adaptive optics.

– *Scanning unit*: super-resolved images are acquired by scanning the spatially and temporally aligned excitation and depletion beams across the sample in unison. Different scanning techniques can be used for this purpose such as beam scanning, stage scanning and a combination of both, each with its own set of strengths and weaknesses. Beam scanning with galvanometer-driven mirrors enables versatile and fast scanning over large fields of view. However, this technique necessitates precise optical alignment and conjugation of the scan unit with the objective, where even minor misalignment results in aberrations. Stage scanning, where the sample is moved via a piezo stage, is very precise, optically simpler and more stable. On the downside, it is relatively slow and difficult to mount micromanipulators for electrophysiological recordings, which must remain stationary in relation to the sample throughout scanning. In a hybrid version, the beam is moved fast along one axis by a resonant mirror (e.g. 15 kHz), while the sample is moved more slowly along the perpendicular axis by the piezo stage. In our implementation, we used a fast telecentric scanner to acquire super-resolved STED images.

– *Image acquisition*: in addition to scanning schemes, the required spatial and temporal resolution of the STED image limits the acquisition, as discussed in detail in the following sections.

i) *Spatial resolution*: an intrinsic limitation on this is imposed by the Nyquist sampling theorem, which specifies that images should be acquired with a pixel size of less than half the optical resolution. As the optical resolution increases, more pixels must be scanned, resulting in slower image acquisition. For instance, pixel sizes of 30 nm or less are needed for an optical resolution of 60 nm.

ii) *Temporal resolution*: the amount of time spent at each pixel (pixel dwell time) determines the acquisition speed, which is ultimately dictated by fluorophore brightness and the speed of the scanning unit. For very bright fluorophores, the scan speed becomes limiting. Typically, pixel dwell times in STED are on the order of tens of microseconds. To increase the temporal resolution, various approaches can be applied. For example, using a stage-scanning system, acquisition times of roughly 10 s per frame were used for imaging volume-labeled neuronal morphology over a 25-μm field of view (Nägerl et al. 2008). STED imaging of synaptic vesicles at a video rate could be done on small scan areas (2.5 μm × 1.8 μm) using a faster scanning technique based on a resonant mirror and a bright organic dye (Westphal et al. 2008). A parallel STED approach was demonstrated to increase the temporal resolution by a factor of four, where the original excitation/depletion laser beams were split into four pairs to scan the sample (Bingen et al. 2011). Currently, most STED microscopes use two-axis galvo scanners (Tonnesen et al. 2011).

– *Detection*: as only a low number of fluorophores emit photons from the reduced focal volume in STED, the detectors must be very sensitive. Single-photon counting avalanche photodiodes (APDs) or photomultiplier tubes (PMT) are best suited for this purpose. The emitted signal is collected by the same objective that focuses the laser beams onto the sample and de-scanned by traversing the beam scanning unit in the reverse direction. The fluorescence signal is isolated from the depletion beam by appropriate filters before being coupled to the multimode optical fiber with a diameter of ~ 62 μm (0.7 Airy unit). In this configuration, the fiber serves as a confocal pinhole, with a diameter equal to or less than one Airy unit of the microscope. The Airy unit of the microscope can be calculated using the following formula:

$$\text{Airy unit} = (1.22 * \text{excitation wavelength} * \text{total magnification}) / \text{NA} \quad [8.5]$$

The total magnification is the sum of all magnifications involved in the emission light path up to the APD detector. In our implementation, the Airy unit of the microscope is ~ 86 μm, with excitation wavelength 485 nm, total magnification 189 and NA 1.3 (63X Gly objective). It is preferable to use a multimode fiber diameter close to one or, even better, 0.7 Airy unit to have the best optical sectioning. However, because the gain in resolution results from the interaction of the laser beams in the focal plane with the fluorophore, a pinhole is theoretically not required in STED microscopy. At the end of the multimode fiber, a fiber-coupled avalanche photodiode working in photon-counting mode detects the fluorescence. Standard data acquisition cards can be used to synchronize the pulsed laser, scanner, objective piezo z-nanopositioner and light detectors.

Instruments	Product name and company
Microscope body	Inverted: DMI 6000, from Leica Upright: BX51, from Olympus
Objectives	Slice imaging: HCX PL APO 100X/1.47 OIL CS or 63X/1.30 GLYC CORR CS, Leica Microsystems Living mice imaging: Olympus LUMFI 60X, NA 1.1, WATER
Sample stage	380FM, Luigs & Neumann
Excitation laser	Single photon: PDL 800-D, Picoquant, Berlin, Germany, 80 MHz repetition rate, 90 ps pluses Two-photon: femtosecond mode-locked titanium: sapphire laser Tsunami, Spectra-Physics, 80 MHz repetition rate, 100 fs pulses
Depletion laser	Classical: Ti:sapphire laser (MaiTai, Spectra-Physics), 80 MHz repetition rate Updated: Katana 06 HP, NKT Photonics, tunable 20–80 MHz
Phase plates	0–2π mask, RP photonics and 0–π mask, homemade
Scanner	Yanus IV, TILL Photonics
3D SLM module	Abberior Instruments, Germany
Detector	Photodiode: PDA8A2, Thorlabs PMT: MD963, Excelitas Avalanche photodiode: SPCM-AQR-13-FC, PerkinElmer
Data acquisition card	PCIe-6259, National Instruments
Pulse delay box	Tombak from Alphanov or SpectraPhysics, Model3930 Lok-to-Clock

Table 8.1. *Instruments, product and company names*

– *Software*: any scanning software can be used for data acquisition and instrument control of a STED setup because scanning and image acquisition in

STED microscopy are nearly identical to those used in traditional laser-scanning microscopy. There are several commercial and free software (Imspector, ScanImage, ImSwich) that can be used for STED instrumentation and acquisition. Abberior's Imspector is a software designed specifically for STED data acquisition and instrument control. ScanImage (Pologruto et al. 2003) is a publicly available MATLAB package for confocal and two-photon imaging that can be adopted for STED imaging. ImSwich (Moreno et al. 2021) is a free STED imaging software written in Python.

In contrast, there are no specific criteria for analyzing STED images, and we can turn to industry-standard image analysis tools (ImageJ, MATLAB, Python, Imaris, etc.). We developed a software tool (SpineJ) as a semi-automated ImageJ plugin for the quantitative analysis of nanoscale spine morphology (Levet et al. 2020).

Finally, just like traditional confocal images, deconvolution methods can further improve STED images. However, unlike super-resolution methods based on single-molecule localization, STED does not require image processing to produce super-resolved images.

8.3.1. *Labeling and optimization of STED*

– *Labeling strategies*: choosing dyes with appropriate photophysical properties that can be targeted to particular organelles or proteins of interest within cells is a key aspect of STED, like for all SRM techniques (Walker 2009). Several labeling strategies exist that are based on transgenics, physical loading or chemical tricks. The selection of the strategy depends on factors such as live-cell compatibility, dye selection, specificity and generalizability. Immunohistochemistry with photostable and bright organic dyes (such as Atto647N and Chromeo494) conjugated to secondary antibodies is the traditional method for achieving strong and specific labeling. This method, however, cannot be used to study living cells because it requires chemical fixation (and membrane permeabilization in the case of intracellular targets). Other labeling strategies that are compatible with STED imaging have been used successfully in live preparations. Transgenic expression of fluorescent proteins (such as GFP, YFP and variants) was used in live cells to image subcellular organelles (Hein et al. 2008) or neuronal morphology (Nägerl et al. 2008; Tønnesen et al. 2014) with a resolution better than 100 nm. Furthermore, organic dyes can be introduced directly into single cells using patch pipettes, as demonstrated by STED imaging of volume-labeled dendritic spines in brain slices using popular Alexa dyes (Ding et al. 2009; Tønnesen et al. 2011). Finally, the SNAP-tag and HaloTags methods can be used to label genetically encoded fusion proteins with exogenous organic dyes for STED imaging in living cells (Hein et al. 2010; Masch et al. 2018; Erdmann et al. 2019).

– *Evaluation of microscope performance*: before using the microscope for biological studies, the performance of the STED system must be rigorously assessed using two closely related parameters: fluorescence quenching and resolution. Fluorescence quenching indicates how effectively the system can deplete the excited state and send the excited molecules to the ground state, and whether it is necessary to tune the STED pulse delay relative to the excitation pulse. Following the spatial alignment of 3D-STED and excitation beam, a sample with a sea of fluorescence (imaging chamber holding a drop of fluorescent liquid) is imaged to optimize fluorescence quenching. The quenching efficiency can be calculated using the fluorescence signal with and without the 3D-STED beam. Although this process is required (for temporal alignment), but it is insufficient by itself because fine spatial misalignments in excitation and STED beams can result in a low SNR despite good quenching. The resolving power of the microscope must be evaluated by imaging sub-diffraction-sized fluorescent beads that provide a readout for resolution in all three spatial dimensions. 40-nm fluorescent beads imaged in confocal, and 3D-STED should appear significantly smaller in all three dimensions after turning on the STED laser (see Figure 8.2(c)).

– *Multi-color imaging*: multi-color STED imaging is helpful for studying synapse biology since synapses are made up of several closely touching compartments that are surrounded by glial processes normally too small to be resolved by traditional microscopy methods. To study the topographical arrangement and dynamics of these structures, it is important to image them at the same time. Multi-color imaging in STED is more difficult to accomplish than in conventional imaging methods since it requires two distinct lasers (one for excitation and the other for depletion). The high intensities of the STED beam, 10^3 times higher than the excitation, pose a threat to fluorophores. Even a 5% absorption of any dye at the STED wavelength results in photobleaching. The choice of fluorophores, as well as the laser lines for excitation and STED, is critical in multicolor imaging. However, several good solutions exist for multicolor imaging with STED, depending on setup costs, fluorophore options and the complexity of the STED system.

The most obvious approach is the introduction of a separate pair of STED and excitation lasers for visible and far-red fluorophores. However, this approach requires sequential imaging starting with the red channel, which would otherwise be bleached immediately by the excitation from the green STED laser (Donnert et al. 2006). Subsequently, a simpler approach was developed by combining long Stokes shift dyes with a conventional dye, allowing for a single STED laser with multiple excitation lasers (Bottanelli et al. 2016; Sidenstein et al. 2016). Nevertheless, multiple laser techniques require complex beam alignment and chromatic aberration correction, limiting the accuracy of co-localization of the two channels.

Alternative methods have been devised to use only a single set of STED and excitation laser, and these methods use various strategies to implement multicolor imaging: i) separation by fluorescence lifetime (Bückers et al. 2011); ii) a pair of photo-switchable fluorescent proteins that can be switched on and off by blue laser flashes (405 nm) (Willig et al. 2011); and iii) transient binding through DNA sequence (Beater et al. 2015). These techniques, however, fall short of capturing the dynamics of living cells. Alternatively, spectrally similar dyes (EGFP and EYFP) are used for simultaneous live-cell imaging, and the emitted signals from the two dyes are separated using spectral detection and linear unmixing (Tønnesen et al. 2011; Arizono et al. 2020) (see Figure 8.4(c)). This technique was recently combined with the reversibly switchable fluorescent protein (RSFP) for three-color imaging (Willig et al. 2021).

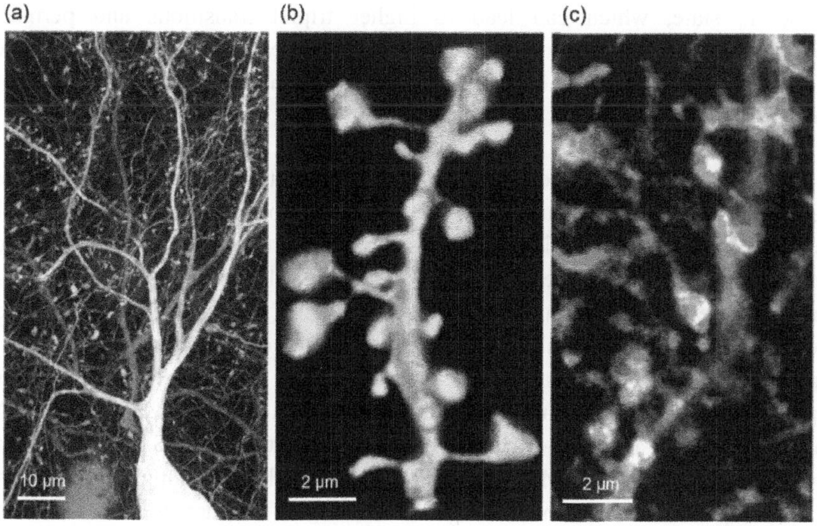

Figure 8.4. *Examples of STED imaging of synapses in living brain slices. (a) A typical pyramidal neuron in the brain, which projects elaborate dendritic and axonal arbors. (b) Three-dimensional reconstruction of a stretch of the dendrite, imaged with 3D STED. (c) Two-color 3D STED of astrocytes (green) and dendrites (magenta) using GFP and YFP as volume markers. (c) Reprinted from Arizono et al. (2020)*

– *Photobleaching*: this is one of the major concerns in light microscopy due to the transition of fluorescent dye to higher-lying reactive states, such as triplet states and their progenies. These states bring temporary bleaching, as well as permanent bleaching due to their high risk of entering into an irreversible photochemical reaction (Song et al. 1996). In principle, two bleaching pathways can be imaginable in STED (see Figure 8.5). In the first path, where the absorption of a fluorescent-state molecule (S_1) leads to a higher molecular singlet state (S_1 to $S_{x>1}$), as proposed for multiphoton microscopy (Dyba and Hell 2003), and these super-excited molecules may bleach through triplet states or return to the S_1 state. However, this path (multiphoton absorption) is almost negligible with the long picosecond pulses. Moreover, as the STED laser is tailored to quench molecules from S_1 to S_0 state, the excitation (S_1 to $S_{x>1}$) by the STED laser is less probable. In the second path, a few molecules in the S_1 state (1–10% per excitation cycle) can spontaneously transfer to the T_1 state via intersystem crossing (ISC), where the stimulated emission does not override the ISC. The molecules accumulated at the T_1 state cause temporary bleaching as they take a long time (0.5–3μs) to relax to the S_0 state (see Figure 8.5). However, if the molecules interact with the subsequent STED pulses while they are in the T_1 state, which can lead to higher triplet transitions and permanent photobleaching. This is the major source of photobleaching in STED microscopy since high intensities and longer dwell times permanently bleach most of the molecules in the T_1 state.

Different techniques are used to alleviate photobleaching via optical and chemical approaches (Li et al. 2018). The optical approaches to resolve this problem are based on fast laser scanning with a resonant mirror or an electro-optical deflector (Borlinghaus 2006; Schneider et al. 2015; Wu et al. 2015), decreasing the laser pulse repetition rate (Donnert et al. 2007), and minimizing the unnecessary exposure to intense light (Göttfert et al. 2017). Fast scanning can successfully reduce the population of triplet states, and as a result, it reduces photobleaching. However, it requires acquiring multiple images from the same area to compensate for low SNR. On the other hand, a similar effect can be achieved by the use of STED lasers with a low-repetition rate (1 < MHz), giving the fluorophores enough time between successive pulses to relax from the triplet states, which can increase the fluorescence yield by 30-fold (Donnert et al. 2006). A downside is that low repetition rate excitation results in slow imaging speed.

The chemical approaches to reduce photobleaching are based on quenching the triplet states, reducing the concentration of free oxygen to minimize destructive chemical reactions, and development of new fluorophores. We have developed an "inverted" labeling strategy in combination with 3D-STED to mitigate photobleaching and phototoxicity, discussed in the section on the recent super-resolution shadow imaging technique.

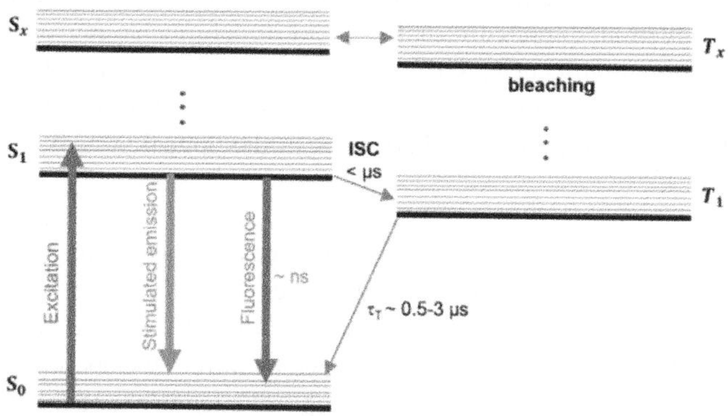

Figure 8.5. *Jablonski energy level diagram and potential photobleaching pathways of fluorophore. ISC: intersystem crossing. For a color version of this figure, see www.iste.co.uk/sibarita/photonic.zip*

8.4. New developments

8.4.1. *Adaptive optics for 3D-STED microscopy*

Many important neurobiological questions require thick tissue samples such as brain slices or even intact brains. However, the achievable spatial resolution and SNR degrade with imaging depth due to aberrations, scattering and absorption induced along the optical path and by the sample. Sample-induced aberrations are mostly spherical and play a critical role in beam shape and null purity, becoming appreciable when imaging not just close to the surface of the tissue (> 10 µm). Nonetheless, alternative ways are being developed to image deeper into thick samples with 2D-STED. For example, using a glycerol objective with an aberration-reducing correction ring, we demonstrated the ability to image synapses inside living tissue close to 100 µm inside the tissue with a 60-nm spatial resolution (Urban et al. 2011). Combining two-photon excitation with STED is another strategy for improving deep tissue imaging (Bethge et al. 2013; Takasaki et al. 2013) two-photon improves depth penetration and optical sectioning. However, while imaging deep, these techniques can maintain a lateral resolution of approximately 60 nm, but the axial resolution is quite poor, calling for 3D-STED.

The fundamental difficulty in 3D-STED imaging of thick samples is that the bottle beam is extremely sensitive to aberrations, much more so than the doughnut beam. As a result, even 2D resolution and SNR decline in 3D-STED due to nonuniform intensity distribution and residual intensity in the null. However, aberrations can be corrected for by approaches based on adaptive optics (AO) where the aberrated wavefront is pre-corrected by SLM or deformable mirror (Booth et al. 2015). SLMs offer significant versatility and can be used for a variety of purposes, including shaping the depletion beam 3D PSFs as discussed in the previous section and, more critically, correcting for aberrations in the optical system as well as the sample.

There are two commonly used AO techniques for assessing and correcting aberrations: closed-loop (with sensor) and open-loop (without sensor). In the closed-loop technique, aberrations are measured directly with a Shack–Hartman wavefront sensor and fed back to the AO device to correct the aberrations (Velasco et al. 2021). This technique requires complex optical alignment and a reference point (guide star) in the sample to measure and correct aberrations. To address these issues, sensorless AO systems were developed. These approaches optimize image quality (intensity and/or sharpness) through the use of a series of predetermined bias aberrations (Gould et al. 2012; Lenz et al. 2014; Patton et al. 2016). This method simplifies a complex optical alignment and does not require a guide star. As a drawback, it requires the acquisition of numerous images to assess and correct aberrations, which can cause additional bleaching and phototoxicity.

We adopted a sensorless AO to image acute brain slices in depth (~ 100 μm) with two-photon 3D-STED, but instead of repeated imaging on biological samples, we used a phantom sample to correct most of the spherical aberrations a priori. The phantom sample is prepared with an agarose gel containing gold and fluorescent beads, and its refractive index is closely adjusted to that of living brain tissue. Before beginning the biological experiments, we imaged the phantom sample to correct the aberrations in depth to preserve the 3D-STED beam shape and the purity of the null (see Figure 8.6(a)). These calibrated values are applied on the AO device to correct aberrations at various imaging depths of the biological sample. We used an SLM to generate 3D-STED PSFs as well to correct the system and sample aberrations.

Images of dendritic spines were acquired at various depths in 2P, uncorrected and corrected STED modes, demonstrating the gain in resolution and signal strength obtained with the AO correction (see Figure 8.6(b)) (Bancelin et al. 2021). Uncorrected STED resolution decreased rapidly with depth, eventually becoming worse than 2P beyond 70 μm. With AO correction, the average spine neck width measured approximately ~ 160 ±15 nm (see Figure 8.6(c)). At a depth of 90 μm, the smallest measurable neck widths were 81 ± 4 nm, which is well below the diffraction limit, providing an upper bound of the effective STED resolution at this depth after correction. Due to the poor SNR, estimating the same neck width in the STED images acquired without correction was impossible. This echoes the issue with the repeated imaging strategy to correct aberrations, where the sample may bleach before suitable aberration parameters are determined, resulting in poor SNR.

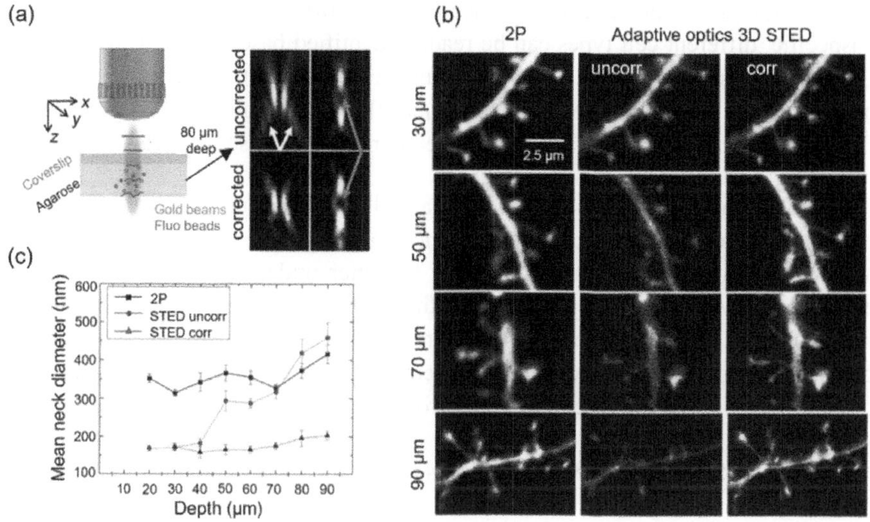

Figure 8.6. *(a) Schematic of the phantom sample imaging, doughnut and bottle beams side view, visualized using gold beads at an 80-µm depth without and with correction. (b) Images of dendritic segments of YFP-labeled pyramidal neurons in hippocampal acute slices at various depth; (c) average spine neck diameter measured in 2P, uncorrected and corrected 3D STED modes. All images are reprinted from Bancelin et al. (2021). For a color version of this figure, see www.iste.co.uk/sibarita/photonic.zip*

8.4.2. *Super-resolution shadow imaging (SUSHI)*

Despite its many strengths, fluorescence microscopy has the disadvantage of only revealing labeled structures, keeping the rest literally in the dark. To provide a much more unbiased and comprehensive view, we devised an "inverted contrast" strategy for visualizing neuroanatomical structures in brain tissue. Instead of labeling individual cells, we label the spaces between them using a highly diffusible yet membrane-impermeable fluorescent dye and image them with a 3D-STED microscope. These in-between spaces are known collectively as the extracellular space of the brain (ECS), which accounts for approximately 20% of the volume of the brain and contains cerebrospinal fluid and the extracellular matrix (Syková and Nicholson 2008).

Since all cells appear as dark shadows in a bright sea of fluorescence, we termed the technique "super-resolution shadow imaging", or SUSHI (Tønnesen et al. 2018). SUSHI generates a super-resolved negative imprint of the space occupied by membrane-bound cellular structures (see Figure 8.7(a)). Even though the labeling is unspecific, different cell types can be readily identified based on their morphology. Due to the nanoscale spatial resolution and favorable contrast settings, it is even possible to see synaptic clefts, where the super-thin yet brightly marked extracellular fissures stand out against the pitch-black cellular structures (see Figure 8.7(a)).

Labeling the ECS rather than individual cells has various practical advantages, including the ease of application, resistance to photobleaching and phototoxicity, and multicolor imaging, which allows us to better understand the interactions of certain cell types with the ECS and other cells at the nanoscale. Because of the diffusional replenishment from a practically infinite pool of fluorophores, photobleaching is not an issue, and any potential phototoxic molecules created outside of the cells are rapidly washed away. As proof, hundreds of time-lapse images can be acquired with SUSHI with no noticeable photodynamic side effects. Simultaneous imaging of ECS with positively labeled cells aids in revealing anatomical relationships from the macro-down to the nanoscale (see Figure 8.7(a) (right)).

8.4.3. *Correlative imaging with STED and SMLM*

Many important biological questions require access to both morphological and molecular information at the nanoscale. For a comprehensive view of synapses, it is necessary to visualize all morphological partners (pre, post and glial cells) as well as synaptic proteins to better understand their roles in synapse structure and function. However, most current SR microscopes do not offer a good way to reveal both protein distributions and the morphological shape of cells. Recently, hybrid SRM modalities were introduced that incorporate optical motifs of STED to enhance the spatial precision of SMLM, such as MINFLUX (Balzarotti et al. 2017) and LocSTED (Puthukodan et al. 2020), but they remain essentially single-molecule imaging techniques.

We have recently overcome this problem by combining STED with SMLM techniques such as (spt)PALM and uPAINT, on a single platform to access morphological and molecular information (Inavalli et al. 2019). It places the molecular information in the context of the morphological organization of the cells, which is crucial for deciphering cellular nanobiology. In essence, the new platform makes it possible to closely correlate, in space and time, the data from both imaging modalities. Based on an optimized workflow, we can rapidly and repeatedly switch back and forth between the SMLM and STED modes.

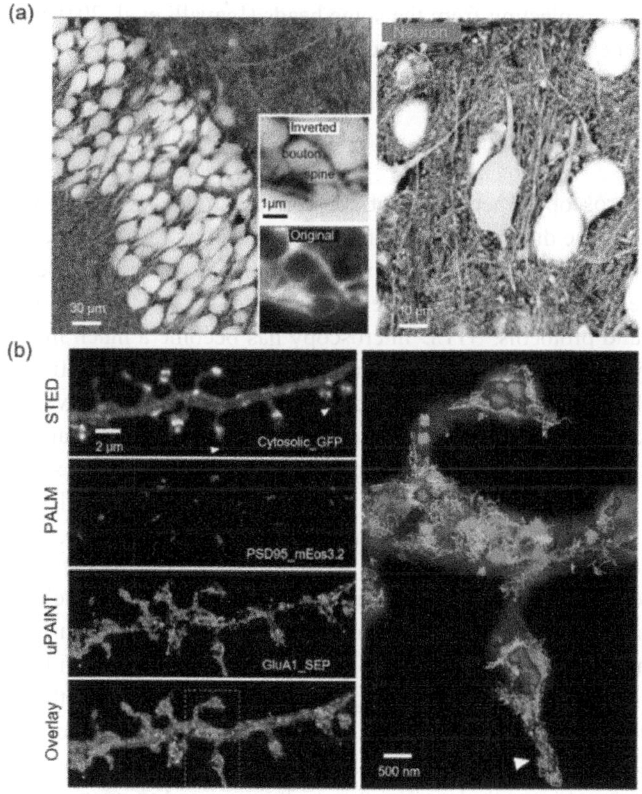

Figure 8.7. *(a) An inverted SUSHI image showing imprints of all brain cells in the tissue. This can be combined with the positive labeling of a particular cell type such as neurons (two-color imaging). (b) The combination of STED and SMLM makes it possible to image nanoscale dendritic spine morphology (orange, STED) together with scaffolding proteins (pink, PALM) and receptor dynamics (blue, uPAINT). The white arrow indicates "spinules", a type of "mini-spine" extending from dendritic spines, where receptors are especially dynamic (right). (a) Reprinted from Tønnesen et al. (2018) and (b) from Inavalli et al. (2019). For a color version of this figure, see www.iste.co.uk/sibarita/photonic.zip*

Using this new approach, we could resolve the distribution and diffusional mobility of several prominent synaptic proteins, such as glutamate receptors (subunits of AMPA receptors, GluA1 and GluA2) and scaffolding proteins (PSD95), within distinct dendritic microstructures in dissociated neuronal cell cultures (see Figure 8.7(b)). We observed that GluA1 molecules are basically arrested inside PSD95 clusters in the spine head but diffuse around rapidly in spinules, which are tiny protrusions projecting from the spine heads (Inavalli et al. 2019), giving us new insights into the nanoscale dynamics of the molecular machinery of synapses.

8.5. Summary and outlook

STED microscopy emerged about 20 years ago, demonstrating for the first time the breaking of the diffraction limit in light microscopy. Since then, the technology has been widely disseminated and is used in many different research fields. In neuroscience, it has had its greatest impact to date. Because of advances in technology and protocols, STED microscopy has become a much more reliable and user-friendly method, which is less complex to implement and difficult to maintain. Commercial STED microscopes are now available from several companies, including Abberior Instruments, Leica Microsystems and Picoquant. Nevertheless, the quest for live-cell nanoscale imaging with more resolution, speed, depth and colors and lower cost, invasiveness and toxicity (as well as less training and frustration) is likely to remain a work in progress for years to come.

To address the current limitations in STED microscopy, researchers are pushing technological boundaries through concurrent developments in various fields. The development of high-power lasers and sensitive SPAD cameras to improve temporal resolution via parallelized scanning schemes, as well as the development of improved switchable nanoprobes and inorganic fluorophores for prolonged time-lapse and multi-color imaging (e.g. with RESOLFT), and the use of adaptive optics in combination with deep-learning techniques, aid in imaging deeper into the brain and achieving isotropic 3D resolution.

In recent years, correlative microscopy approaches have triggered a new frontier in the research field to unravel the complex architecture of neurobiology by taking advantage of distinct imaging techniques. We explored the combination of STED with two single-molecule localization microscopy techniques to unravel morphological and molecular information of neuronal synapses. For a comprehensive view of the brain, we need multiscale imaging modalities to access information from the nano to the millimeter spatial scale at different temporal resolutions. Combination of SRM techniques with label-free imaging modalities such as photoacoustic imaging, MRI and Raman spectroscopy may help fill this gap.

However, implementing these combinations is a daunting task that requires multidisciplinary expertise to overcome the technical challenges.

8.6. References

Arizono, M., Inavalli, V.V.G.K., Panatier, A., Pfeiffer, T., Angibaud, J., Levet, F., Ter Veer, M.J.T., Stobart, J., Bellocchio, L., Mikoshiba, K. et al. (2020). Structural basis of astrocytic Ca2+ signals at tripartite synapses. *Nature Communications 2020*, 11(1), 1–15. doi: 10.1038/s41467-020-15648-4.

Balzarotti, F., Eilers, Y., Gwosch, K.C., Gynnå, A.H., Westphal, V., Stefani, F.D., Elf, J., Hell, S.W. (2017). Nanometer resolution imaging and tracking of fluorescent molecules with minimal photon fluxes. *Science*, 355(6325), 606–612. doi: 10.1126/science.aak9913.

Bancelin, S., Mercier, L., Murana, E., Nägerl, U.V. (2021). Aberration correction in stimulated emission depletion microscopy to increase imaging depth in living brain tissue. *Neurophotonics*, 8(3), 035001. doi: 10.1117/1.NPH.8.3.035001.

Beater, S., Holzmeister, P., Lalkens, B., Tinnefeld, P. (2015). Simple and aberration-free 4color-STED – Multiplexing by transient binding. *Optics Express*, 23(7), 8630. doi: 10.1364/OE.23.008630.

Bethge, P., Chéreau, R., Avignone, E., Marsicano, G., Nägerl, U.V. (2013). Two-photon excitation STED microscopy in two colors in acute brain slices. *Biophysical Journal*, 104(4), 778–785. doi: 10.1016/J.BPJ.2012.12.054.

Betzig, E., Patterson, G.H., Sougrat, R., Lindwasser, O.W., Olenych, S., Bonifacino, J.S., Davidson, M.W., Lippincott-Schwartz, J., Hess, H.F. (2006). Imaging intracellular fluorescent proteins at nanometer resolution. *Science*, 313(5793), 1642–1645. doi: 10.1126/science.1127344.

Bingen, P., Reuss, M., Engelhardt, J., Hell, S.W. (2011). Parallelized STED fluorescence nanoscopy. *Optics Express*, 19(24), 23716–23726. doi: 10.1364/OE.19.023716.

Blom, H. and Widengren, J. (2017). Stimulated emission depletion microscopy. *Chemical Reviews*, 117(11), 7377–7427. doi: 10.1021/acs.chemrev.6b00653.

Booth, M., Andrade, D., Burke, D., Patton, B., Zurauskas, M. (2015). Aberrations and adaptive optics in super-resolution microscopy. *Microscopy*, 64(4), 251–261. doi: 10.1093/jmicro/dfv033.

Borlinghaus, R.T. (2006). MRT letter: High speed scanning has the potential to increase fluorescence yield and to reduce photobleaching. *Microscopy Research and Technique*, 69(9), 689–692. doi: 10.1002/JEMT.20363.

Bottanelli, F., Kromann, E.B., Allgeyer, E.S., Erdmann, R.S., Baguley, S.W., Sirinakis, G., Schepartz, A., Baddeley, D., Toomre, D.K., Rothman, J.E., Bewersdorf, J. (2016). Two-colour live-cell nanoscale imaging of intracellular targets. *Nature Communications 2016*, 7(1), 1–5. doi: 10.1038/ncomms10778.

Brouwer, B. (2016). Polarization effects in super resolution STED microscopy. Bachelor's Thesis, Utrecht University, Utrecht.

Bückers, J., Wildanger, D., Vicidomini, G., Kastrup, L., Hell, S.W. (2011). Simultaneous multi-lifetime multi-color STED imaging for colocalization analyses. *Optics Express*, 19(4), 3130. doi: 10.1364/OE.19.003130.

Chéreau, R., Tønnesen, J., Nägerl, U.V. (2015). STED microscopy for nanoscale imaging in living brain slices. *Methods*, 88, 57–66. doi: 10.1016/J.YMETH.2015.06.006.

Coons, A.H. and Kaplan, M.H. (1950). Localization of antigen in tissue cells: II. improvements in a method for the detection of antigen by means of fluorescent antibody. *The Journal of Experimental Medicine*, 91(1), 1. doi: 10.1084/JEM.91.1.1.

Deng, S., Liu, L., Cheng, Y., Li, R., Xu, Z. (2010). Effects of primary aberrations on the fluorescence depletion patterns of STED microscopy. *Optics Express*, 18(2), 1657. doi: 10.1364/oe.18.001657.

Ding, J.B., Takasaki, K.T. Sabatini, B.L. (2009). Supraresolution imaging in brain slices using stimulated-emission depletion 2-photon laser scanning microscopy. *Neuron*, 63(4), 429. doi: 10.1016/J.NEURON.2009.07.011.

Donnert, G., Keller, J., Medda, R., Andrei, M.A., Rizzoli, S.O., Lührmann, R., Jahn, R., Eggeling, C., Hell, S.W. (2006). Macromolecular-scale resolution in biological fluorescence microscopy. *Proceedings of the National Academy of Sciences of the USA*, 103(31), 11440–11445. doi: 10.1073/pnas.0604965103.

Donnert, G., Eggeling, C., Hell, S.W. (2007). Major signal increase in fluorescence microscopy through dark-state relaxation. *Nature Methods*, 4(1), 81–86. doi: 10.1038/nmeth986.

Dyba, M. and Hell, S.W. (2003). Photostability of a fluorescent marker under pulsed excited-state depletion through stimulated emission. *Applied Optics*, 42(25), 5123–5129. doi: 10.1364/AO.42.005123.

Erdmann, R.S., Baguley, S.W., Richens, J.H., Wissner, R.F., Xi, Z., Allgeyer, E.S., Zhong, S., Thompson, A.D., Lowe, N., Butler, R. et al. (2019). Labeling strategies matter for super-resolution microscopy: A comparison between HaloTags and SNAP-tags. *Cell Chemical Biology*, 26(4), 584–592.e6. doi: 10.1016/J.CHEMBIOL.2019.01.003.

Fürstenberg, A. and Heilemann, M. (2013). Single-molecule localization microscopy – Near-molecular spatial resolution in light microscopy with photoswitchable fluorophores. *Physical Chemistry Chemical Physics*, 15(36), 14919–14930. doi: 10.1039/C3CP52289J.

Giannone, G., Hosy, E., Levet, F., Constals, A., Schulze, K., Sobolevsky, A.I., Rosconi, M.P., Gouaux, E., Tampé, R., Choquet, D., Cognet, L. (2010). Dynamic superresolution imaging of endogenous proteins on living cells at ultra-high density. *Biophysical Journal*, 99(4), 1303–1310. doi: 10.1016/j.bpj.2010.06.005.

Göttfert, F., Pleiner, T., Heine, J., Westphal, V., Görlich, D., Sahl, S.J., Hell, S.W. (2017). Strong signal increase in STED fluorescence microscopy by imaging regions of subdiffraction extent. *Proceedings of the National Academy of Sciences of the USA*, 114(9), 2125–2130. doi: 10.1073/pnas.1621495114.

Gould, T.J., Burke, D., Bewersdorf, J., Booth, M.J. (2012). Adaptive optics enables 3D STED microscopy in aberrating specimens. *Optics Express*, 20(19), 20998. doi: 10.1364/OE.20.020998.

Hao, X., Kuang, C., Wang, T., Liu, X. (2010). Effects of polarization on the de-excitation dark focal spot in STED microscopy. *Journal of Optics*, 12(11), 115707. doi: 10.1088/2040-8978/12/11/115707.

Harke, B., Keller, J., Ullal, C.K., Westphal, V., Schönle, A., Hell, S.W. (2008). Resolution scaling in STED microscopy. *Optics Express*, 16(6), 4154. doi: 10.1364/oe.16.004154.

Hein, B., Willig, K.I., Hell, S.W. (2008). Stimulated emission depletion (STED) nanoscopy of a fluorescent protein-labeled organelle inside a living cell. *Proceedings of the National Academy of Sciences of the USA*, 105(38), 14271–14276. doi: 10.1073/pnas.0807705105.

Hein, B., Willig, K.I., Wurm, C.A., Westphal, V., Jakobs, S., Hell, S.W. (2010). Stimulated emission depletion nanoscopy of living cells using SNAP-tag fusion proteins. *Biophysical Journal*, 98(1), 158–163. doi: 10.1016/J.BPJ.2009.09.053.

Hell, S.W. (2007). Far-field optical nanoscopy. *Science*, 316(5828), 1153–1158. doi: 10.1126/science.1137395.

Hell, S.W. and Wichmann, J. (1994). Breaking the diffraction resolution limit by stimulated emission: Stimulated-emission-depletion fluorescence microscopy. *Optics Letters*, 19(11), 780. doi: 10.1364/ol.19.000780.

Inavalli, V.V.G.K., Lenz, M.O., Butler, C., Angibaud, J., Compans, B., Levet, F., Tønnesen, J., Rossier, O., Giannone, G., Thoumine, O. et al. (2019). A super-resolution platform for correlative live single-molecule imaging and STED microscopy. *Nature Methods*, 16(12). doi: 10.1038/s41592-019-0611-8.

Klar, T.A. and Hell, S.W. (1999). Subdiffraction resolution in far-field fluorescence microscopy. *Optics Letters*, 24(14), 954. doi: 10.1364/OL.24.000954.

Kubitscheck, U. (2017). *Fluorescence Microscopy: From Principles to Biological Applications*, 2nd edition. John Wiley & Sons, New York. doi: 10.1002/9783527687732.

Lelek, M., Gyparaki, M.T., Beliu, G., Schueder, F., Griffié, J., Manley, S., Jungmann, R., Sauer, M., Lakadamyali, M., Zimmer, C. (2021). Single-molecule localization microscopy. *Nature Reviews Methods Primers*, 1(1), 39. doi: 10.1038/s43586-021-00038-x.

Lenz, M.O., Sinclair, H.G., Savell, A., Clegg, J.H., Brown, A.C.N., Davis, D.M., Dunsby, C., Neil, M.A.A., French, P.M.W. (2014). 3-D stimulated emission depletion microscopy with programmable aberration correction. *Journal of Biophotonics*, 7(1–2), 29–36. doi: 10.1002/JBIO.201300041.

Levet, F., Tønnesen, J., Nägerl , U.V., Sibarita, J.-B. (2020). SpineJ: A software tool for quantitative analysis of nanoscale spine morphology. *Methods*, 174, 49–55. doi: 10.1016/j.ymeth.2020.01.020.

Li, C., Liu, S., Wang, W., Liu, W., Kuang, C., Liu, X. (2018). Recent research on stimulated emission depletion microscopy for reducing photobleaching. *Journal of Microscopy*, 271(1), 4–16. doi: 10.1111/JMI.12698.

Masch, J.M., Steffens, H., Fischer, J., Engelhardt, J., Hubrich, J., Keller-Findeisen, J., D'Este, E., Urban, N.T., Grant, S.G.N., Sahl, S.J. (2018). Robust nanoscopy of a synaptic protein in living mice by organic-fluorophore labeling. *Proceedings of the National Academy of Sciences of the USA*, 115(34), E8047–E8056. doi: 10.1073/pnas.1807104115.

Masters, B.R. (2015). Super-resolution microscopy techniques in the neurosciences. *Journal of Biomedical Optics*, 20(1), 019901. doi: 10.1117/1.jbo.20.1.019901.

Moreno, X.C., Al-Kadhimi, S., Alvelid, J., Bodén, A., Testa, I. (2021). ImSwitch: Generalizing microscope control in Python. *Journal of Open Source Software*, 6(64), 3394. doi: 10.21105/JOSS.03394.

Nägerl, U.V., Willig, K.I., Hein, B., Hell, S.W., and Bonhoeffer, T. (2008). Live-cell imaging of dendritic spines by STED microscopy. *Proceedings of the National Academy of Sciences of the USA*, 105(48), 18982–18987. doi: 10.1073/pnas.0810028105.

Patton, B.R., Burke, D., Owald, D., Gould, T.J., Bewersdorf, J., Booth, M.J. (2016). Three-dimensional STED microscopy of aberrating tissue using dual adaptive optics. *Optics Express*, 24(8), 8862. doi: 10.1364/OE.24.008862.

Pawley, J.B. (2006). *Handbook of Biological Confocal Microscopy*, 3rd edition. Springer, New York. doi: 10.1007/978-0-387-45524-2/COVER.

Pologruto, T.A., Sabatini, B.L., Svoboda, K. (2003). ScanImage: Flexible software for operating laser scanning microscopes. *BioMedical Engineering Online*, 2(1), 1–9. doi: 10.1186/1475-925X-2-13/FIGURES/4.

Puthukodan, S., Murtezi, E., Jacak, J., Klar, T. (2020). Localization STED (LocSTED) microscopy with 15 nm resolution. *Nanophotonics*, 9(4), 783–792. doi: 10.1515/nanoph-2019-0398.

Rittweger, E., Han, K.Y., Irvine, S.E., Eggeling, C., Hell, S.W. (2009). STED microscopy reveals crystal colour centres with nanometric resolution. *Nature Photonics*, 3(3), 144–147. doi: 10.1038/NPHOTON.2009.2.

Rust, M.J., Bates, M., Zhuang, X. (2006). Sub-diffraction-limit imaging by stochastic optical reconstruction microscopy (STORM). *Nature Methods*, 3, 793–795. doi: 10.1038/nmeth929.

Sanderson, J. (ed.) (2019). Multi-photon microscopy. In *Understanding Light Microscopy*. John Wiley & Sons, New York. doi: 10.1002/9781118696736.ch22.

Schneider, J., Zahn, J., Maglione, M., Sigrist, S.J., Marquard, J., Chojnacki, J., Kräusslich, H.-G., Sahl, S.J., Engelhardt, J., Hell, S.W. (2015). Ultrafast, temporally stochastic STED nanoscopy of millisecond dynamics. *Nature Methods 2015*, 12(9), 827–830. doi: 10.1038/nmeth.3481.

Sidenstein, S.C., D'Este, E., Böhm, M., Danzl, J.G., Belov, V.N., Hell, S.W. (2016). Multicolour multilevel STED nanoscopy of actin/spectrin organization at synapses. *Scientific Reports*, 6, 26725. doi: 10.1038/SREP26725.

Song, L., Varma, C.A., Verhoeven, J.W., Tanke, H.J. (1996). Influence of the triplet excited state on the photobleaching kinetics of fluorescein in microscopy. *Biophysical Journal*, 70(6), 2959–2968. doi: 10.1016/S0006-3495(96)79866-1.

Syková, E. and Nicholson, C. (2008). Diffusion in brain extracellular space. *Physiological Reviews*, 88(4), 1277–1340. doi: 10.1152/physrev.00027.2007.

Takasaki, K.T., Ding, J.B., Sabatini, B.L. (2013). Live-cell superresolution imaging by pulsed STED two-photon excitation microscopy. *Biophysical Journal*, 104(4), 770. doi: 10.1016/J.BPJ.2012.12.053.

Tønnesen, J., Nadrigny, F., Willig, K.I., Wedlich-Söldner, R., Nägerl, U.V. (2011). Two-color STED microscopy of living synapses using a single laser-beam pair. *Biophysical Journal*, 101(10), 2545–2552. doi: 10.1016/J.BPJ.2011.10.011.

Tønnesen, J., Katona, G., Rózsa, B., Nägerl, U.V. (2014). Spine neck plasticity regulates compartmentalization of synapses. *Nature Neuroscience 2014*, 17(5), 678–685. doi: 10.1038/nn.3682.

Tønnesen, J., Inavalli, V.V.G.K., Nägerl, U.V. (2018). Super-resolution imaging of the extracellular space in living brain tissue. *Cell*, 172(5), 1108–1121.e15. doi: 10.1016/j.cell.2018.02.007.

Urban, N.T., Willig, K.I., Hell, S.W., Nägerl, U.V. (2011). STED nanoscopy of actin dynamics in synapses deep inside living brain slices. *Biophysical Journal*, 101(5), 1277–1284. doi: 10.1016/j.bpj.2011.07.027.

Velasco, M.G.M., Zhang, M., Antonello, J., Yuan, P., Allgeyer, E.S., May, D., M'Saad, O., Kidd, P., Barentine, A.E.S., Greco, V. et al. (2021). 3D super-resolution deep-tissue imaging in living mice. *Optica*, 8(4), 442. doi: 10.1364/optica.416841.

Walker, J.M. (2009). *Live Cell Imaging: Methods and Protocols*. Humana Press, Clifton.

Westphal, V., Rizzoli, S.O., Lauterbach, M.A., Kamin, D., Jahn, R., Hell, S.W. (2008). Video-rate far-field optical nanoscopy dissects synaptic vesicle movement. *Science*, 320(5873), 246–249. doi: 10.1126/science.1154228.

Wildanger, D., Medda, R., Kastrup, L., Hell, S.W. (2009). A compact STED microscope providing 3D nanoscale resolution. *Journal of Microscopy*, 236(1), 35–43. doi: 10.1111/j.1365-2818.2009.03188.x.

Willig, K.I., Stiel, A.C., Brakemann, T., Jakobs, S., Hell, S.W. (2011). Dual-label STED nanoscopy of living cells using photochromism. *Nano Letters*, 11(9), 3970–3973. doi: 10.1021/NL202290W.

Willig, K.I., Wegner, W., Müller, A., Calvet-Fournier, V., Steffens, H. (2021). Multi-label in vivo STED microscopy by parallelized switching of reversibly switchable fluorescent proteins. *Cell Reports*, 35(9), 109192. doi: 10.1016/j.celrep.2021.109192.

Wu, Y., Wu, X., Lu, R., Zhang, J., Toro, L., Stefani, E. (2015). Resonant scanning with large field of view reduces photobleaching and enhances fluorescence yield in STED microscopy. *Scientific Reports 2015*, 5(1), 1–12. doi: 10.1038/srep14766.

9

Single-Molecule Localization Microscopy: From Imaging Cellular Structures to Quantitative Image Analysis

Marina S. DIETZ and Mike HEILEMANN

Single-Molecule Biophysics, Institute of Physical and Theoretical Chemistry, Goethe University Frankfurt, Germany

Optical super-resolution microscopy allows the visualization of cellular structures with a spatial resolution of a few tens of nanometers and has revolutionized our understanding in cell biology. Single-molecule localization microscopy is one powerful realization of super-resolution techniques, which builds on the isolated detection of single fluorophores. This chapter introduces the concept of SMLM and its development into a quantitative, multi-color imaging tool for cell biology.

9.1. Introduction

The spatial resolution in fluorescence microscopy is limited by diffraction (Pawley and Masters 2008). The lateral spatial resolution can be approximated by $\lambda/2$, reaching a lower limit of about 200 nm for blue light. This value is much larger than small structural features in a cell, such as protein assemblies or organelle compartments, which in consequence remain obscured to the direct observation with optical microscopy (see Figure 9.1(A)). In the late 1990s, several experimental concepts that bypass this limit in spatial resolution in far-field light microscopy were

Photonic Imaging for Biology,
coordinated by Jean-Baptiste SIBARITA.

developed (Hell and Wichmann 1994; Betzig 1995; Gustafsson et al. 1995, 2000), of which three turned into major "super-resolution" imaging techniques for optical cell biology now established in many laboratories: stimulated emission depletion (STED) (Sahl and Hell 2019; see also Chapter 8 of this book), structured illumination microscopy (SIM) (Wu and Shroff 2018; see also Chapter 7 of this book) and single-molecule localization microscopy (SMLM) (Sauer and Heilemann 2017). These super-resolution techniques have revolutionized our understanding in cell biology, paving the way to the direct observation of how life is organized in a cell (Sahl et al. 2017).

In this chapter, we will introduce the concept of SMLM and highlight selected applications in cell biology. First, we outline the underlying principle of SMLM and discuss different experimental realizations and fluorophore labels. Next, we discuss the developments towards multi-target labeling and imaging. Lastly, we present concepts of quantitative image data analysis that report on protein numbers and spatial inter-relationships at the nanoscale.

9.2. Concepts for single-molecule localization microscopy

SMLM builds on the separation of fluorescence emission in time and space. Isolated emission profiles (point spread functions, PSFs) of single fluorophores are approximated with a Gaussian function which provides its center of mass (or "localization"). The precision of this approximation scales inversely with the square root of the number of detected photons (Mortensen et al. 2010) and can reach a few nanometers in lateral space. SMLM exploits this nanometer precision to generate a super-resolution image by plotting the positions of all molecules (see Figure 9.1). Various software packages for SMLM analysis are freely available and described in the literature (Wolter et al. 2010; Ovesný et al. 2014; Sage et al. 2019; Schnitzbauer et al. 2017).

SMLM comes in various flavors that can be distinguished by the way that the sparsity of the fluorescence signal is generated (see Figure 9.2). One option is to use fluorophores that can be optically modulated, for example, photoactivatable or photoconvertible fluorescent proteins (Patterson and Lippincott-Schwartz 2002) or photoswitchable organic fluorophores (Heilemann et al. 2005). These fluorophores are used in (fluorescence) photoactivated localization microscopy ((F)PALM) (Betzig et al. 2006; Hess et al. 2006) and (direct) stochastic optical reconstruction microscopy ((d)STORM) (Rust et al. 2006; Heilemann et al. 2008). For the latter method, either quencher fluorophores (STORM) or reducing agents (dSTORM) are used in combination with activation and excitation lasers to switch between a bright and a dark state. To date, a large selection of fluorescent proteins and organic fluorophores is available for SMLM (Fürstenberg and Heilemann 2013; Jradi and Lavis 2019). A second option for SMLM is to use exchangeable fluorophore labels that transiently bind to a target with a weak binding affinity. This was first demonstrated with membrane-intercalating fluorophores in point accumulation for nanoscale topography (PAINT) (Sharonov and Hochstrasser 2006), and later extended, among others, to peptide-based probes (Sharonov and Hochstrasser 2006; Kiuchi et al. 2015) and DNA-intercalating labels (Spahn et al. 2018). PAINT was generalized by implementing the hybridization of short DNA oligonucleotides with weak affinity, in a method termed DNA-PAINT (Jungmann et al. 2010; Steinhauer et al. 2011).

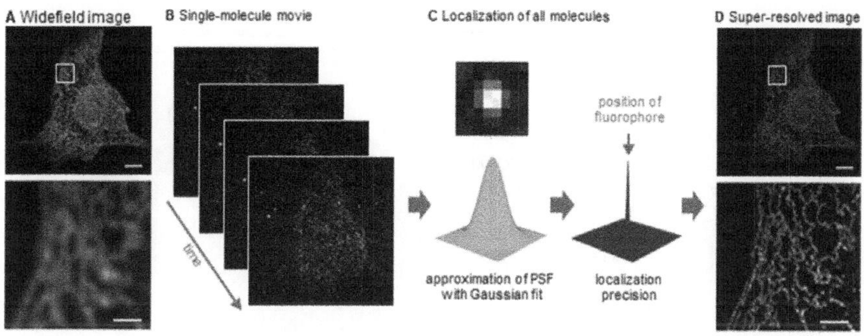

Figure 9.1. *Basic principle of single-molecule localization microscopy. (A) The spatial resolution of widefield fluorescence microscopy is limited by the diffraction of light. Structural features that are smaller than the diffraction limit, such as the fine structure of the endoplasmic reticulum, are not discernible in the widefield image. (B) In SMLM, single-molecule emission events are separated in time and space by recording an image series. (C) Point-spread functions (PSFs) of single fluorophores are approximated with a Gaussian function that reports its center of mass (its "localization"). The precision of the position determination improves with an increasing number of photons. (D) A super-resolved image is reconstructed from all localizations. The tubular structure of the endoplasmic network is discernible in the super-resolved image. Scale bars 10 μm, scale bar in zoom-ins 2 μm. For a color version of this figure, see www.iste.co.uk/sibarita/photonic.zip*

The spatial resolution of an SMLM image can be assessed by different parameters. Specific to SMLM is the precision with which the position of a single fluorophore was determined, which informs on the error in localization. This "localization precision" is derived as the standard deviation of the Gaussian approximation of the detected fluorescence emission spot and scales with 1/sqrt(N), as well as other parameters such as background signal and pixel size (Mortensen et al. 2010). The error in localization can also be determined from a nearest-neighbor analysis, exploiting the fact that most single emitters will be recorded in more than one imaging frame; the distance between the positions of the same fluorophore in neighboring imaging frames reports on the experimental localization error (nearest neighbor-based analysis, NeNA) (Endesfelder et al. 2014). In addition to the localization error, the fluorophore label contributes a linkage error, which represents the distance between the fluorophore and the target molecule. This linkage error depends on the label size and can surpass the localization precision (Moore and Legant 2018).

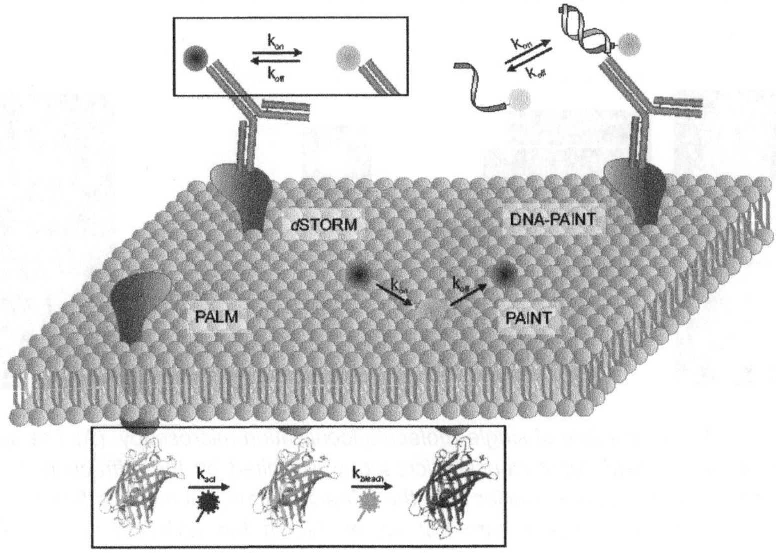

Figure 9.2. *Fluorophore labels for single-molecule localization microscopy. The scheme shows four types of fluorophore labels for SMLM. In PALM, a photoactivatable or photoconvertible fluorescent protein is genetically coupled to a target protein and expressed in a cell. Upon irradiation with, for example, violet light, single fluorescent proteins are stochastically transferred to a fluorescent state which is detectable by excitation with visible light (see the box at the bottom). dSTORM relies on reversibly photoswitchable organic dyes. Photoswitching is induced by reducing buffers and illumination light, and the fluorophores are reversibly switched between a fluorescent and a long-lived dark state (see the box at the top). In PAINT, fluorophores are kept in an imaging buffer and transiently bind to their target (here shown for a fluorogenic, membrane-intercalating fluorophore). Target binding leads to a transient localization and to a detectable fluorescence signal. DNA-PAINT exploits the transient hybridization of short oligonucleotides, with a docking strand attached to the target (or a label that binds to a target) and a sequence-complementary and fluorophore-labeled imager strand diffusing in solution. For a color version of this figure, see www.iste.co.uk/sibarita/photonic.zip*

While the localization error is a measure of the precision of position determination based on the fluorescence signal, it is not per se a measure of the spatial resolution. Another relevant parameter is the labeling density of a structure. Applying the Nyquist–Shannon sampling theorem to SMLM, the labeling density has to be at least twice as high as the desired spatial resolution (Shannon 1984). The influence of the labeling density, together with other parameters such as the photoswitching rates of fluorophores and the structural dimensionality of a sample were investigated in the context of SMLM (van de Linde et al. 2010).

At the end, however, the true spatial resolution of an image is the desired information, considering all the above parameters that arise from an imaging experiment and the reconstruction of an SMLM image. This information can be obtained by either using Fourier Ring Correlation (FRC) (Banterle et al. 2013; Nieuwenhuizen et al. 2013) or image decorrelation analysis (Descloux et al. 2019, 2020, 2021). We note that both methods for resolution determination are straight-forward to use, and free software is available; for a detailed description of these methods, we refer to the original literature.

9.3. Multiplexing

Cellular processes are orchestrated by the interaction of many biomolecules. For example, membrane receptor activation is often initiated by ligand binding, followed by the recruitment of adaptor proteins and the formation of multi-protein "signaling hubs". The information is then "transported" through the cell by a cascade of proteins until ultimately a transcription factor modulates gene expression (Wardhan and Mudgal 2017). In order to understand such complex cellular processes, we need to map the relevant proteins in the same context, i.e. the same cell. Multi-target fluorescence microscopy using high-affinity or covalent fluorophore labels uses, for example, emission wavelength or fluorescence lifetime to distinguish between targets.

However, this limits the observation to a few targets depending on how many distinguishable fluorophores are available. In this section, we introduce some examples that bypass this limitation and enable the study of multi-protein assemblies in cells.

The structure of complex multi-protein assemblies that exhibit a high symmetry can be investigated by sequential imaging of the relative position of a set of proteins with respect to the same reference protein. Using this concept, the organization of the Y-complex in the nuclear pore complex (NPC) (Szymborska et al. 2013), and the protein organization in clathrin-mediated endocytosis (Mund et al. 2018) (see Figure 9.3(A)) were investigated. This overall very powerful approach has two challenges: first, it is limited to symmetrical structures, or at least the assumption of some degree of symmetry; second, the structural information obtained from different samples is averaged, such that information on structural heterogeneity and on rare phenotypes might be lost.

Another experimental strategy for multiplexed SMLM is the implementation of sequential steps of immunolabeling and -unlabeling of the same sample into an imaging experiment, which was demonstrated to largely increase the number of proteins observable in an SMLM experiment (Yi et al. 2016; Klevanski et al. 2020) (see Figure 9.3(B)).

For example, the structural organization of 14 proteins in individual active zones was mapped in semi-thin brain tissue, and protein heterotypes were characterized (Klevanski et al. 2020). This approach removes the need for symmetry and can report on the structural organization of individual protein assemblies.

A further option is to use weak-affinity fluorophore labels that repetitively bind to and unbind from a target molecule (Albertazzi and Heilemann 2023). A powerful realization of this concept is to "program" the target specificity of a label into short DNA oligonucleotides, as in DNA-PAINT (Schnitzbauer et al. 2017). Multiplexing is achieved by using orthogonal oligonucleotides in the same imaging experiment. The orthogonal oligonucleotides are conjugated to target-specific labels, for example, antibodies, and are read-out by complementary, fluorophore-labeled oligonucleotides that are sequentially added to the sample, imaged and washed out again (see Figure 9.3(C)).

This concept of multi-protein imaging (Exchange-PAINT) (Jungmann et al. 2014) was used to study the protein organization of membrane protein networks (Werbin et al. 2017; Harwardt et al. 2020; Schröder et al. 2021) or the adhesion complex (Fischer et al. 2021). A variation of this principle uses complementary oligonucleotides which carry quencher molecules and enabler imaging of thick tissue samples without washing (Lutz et al. 2018; Schueder et al. 2024; Unterauer et al. 2024).

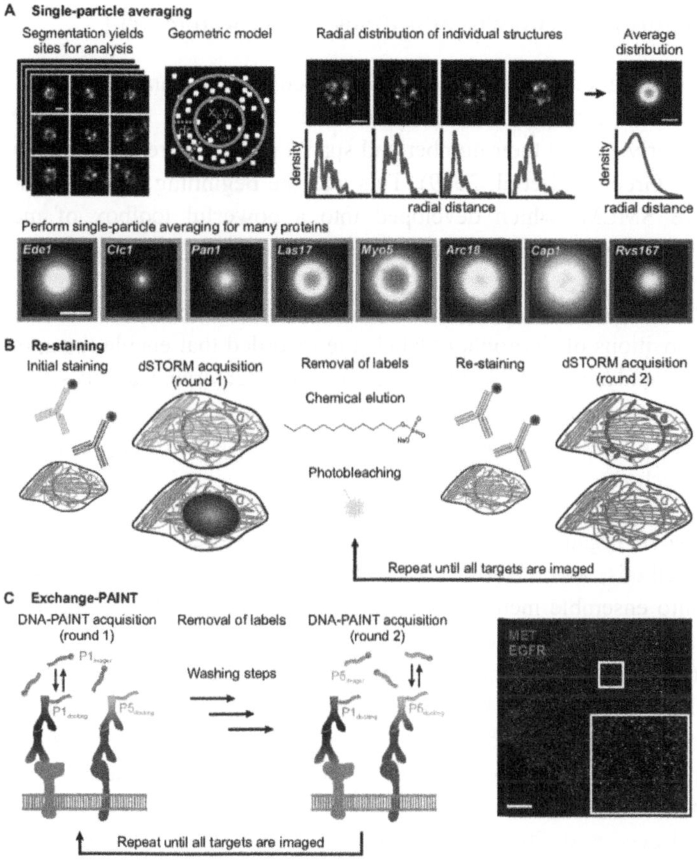

Figure 9.3. *Imaging multiple cellular targets with single-molecule localization microscopy. (A) Single-particle averaging. SMLM images are segmented and fitted to a geometric model, yielding an averaged spatial distribution. By performing single-particle averaging for different proteins within a multiprotein complex, the organization of symmetric structures, such as endocytic sites, can be determined. Scale bars 100 nm. Adapted from Mund et al. (2018). Copyright CC BY 4.0. (B) Repeated cycles of staining and label removal. In a first round, two different structures are labeled with two spectrally distinguishable fluorophores and visualized. Next, the fluorophore labels are removed by either chemical elution or photobleaching, and two new targets are labeled for another round of imaging. This cycle can be repeated multiple times and allows imaging the structural context of multi-protein assemblies in cells and tissue. Adapted from Klevanski et al. (2020). Copyright CC BY 4.0. (C) Exchange-PAINT. Multiplexing is achieved by repeated cycles of DNA-PAINT imaging, adding an imager strand in each imaging cycle that specifically binds to one target. Scale bar 5 µm. Adapted from Harwardt et al. (2020). Copyright CC BY 4.0. For a color version of this figure, see www.iste.co.uk/sibarita/photonic.zip*

9.4. Extracting quantitative information from SMLM data

Early applications of SMLM mostly focused on unraveling the nanoscale organization of proteins in cells. However, individual protein clusters were visualized early on and their number and spatial pattern were studied (van de Linde et al. 2008; Greenfield et al. 2009). This was the beginning of the development of quantitative SMLM, which developed into a powerful toolbox of methods for quantitative cell biology (Nicovich et al. 2017).

Quantitative analysis can profit from the nature of SMLM data: first, single-molecule positions of fluorophore labels are recorded that enable the application of point-based analysis methods to report on molecular organization; second, kinetic information originating from fluorescence photophysics or label binding can be converted to molecule numbers. Both approaches will be briefly introduced in the following section.

The spatial organization of proteins can be captured with point-based cluster analysis methods, and a variety of methods are available to date. These can be grouped into ensemble methods, which average over the entire SMLM image, and those that detect individual clusters. Ensemble cluster analysis includes Ripley's K function (Ripley 1976) and pair-correlation analysis (Sengupta et al. 2011; Sengupta and Lippincott-Schwartz 2012). Both approaches yield information on the average length scale of clustering and the cluster density. Individual cluster analysis methods include point-based density analyses (Ester et al. 1996; Ankerst et al. 1999; Campello et al. 2013) and the generation of mesh representations such as Voronoi diagrams (Levet et al. 2015; Andronov et al. 2016). These methods inform on the size, density and molecular numbers for each individual cluster. Cluster analysis can be extended to multiple species, which is relevant if the local co-organization or interaction of two (or more) proteins is questioned. This can be addressed with colocalization analysis, either by analyzing single-molecule localization patterns of two proteins (Malkusch et al. 2012), by generating localization clusters (Levet et al. 2019) or by a combination of both (Pageon et al. 2016).

Assessing the spatial organization of proteins within densely packed clusters remains challenging if the attainable spatial resolution is insufficient to resolve individual components. This is, for example, the case for membrane receptors that often assemble into multimers to build functional units (Karathanasis et al. 2020). In order to extract quantitative information from such protein nano-clusters, the analysis of fluorescence emission kinetics originating either from photoswitching cycles or repetitive binding events was implemented into SMLM (Lee et al. 2012; Ehmann et al. 2014; Fricke et al. 2015; Jungmann et al. 2016).

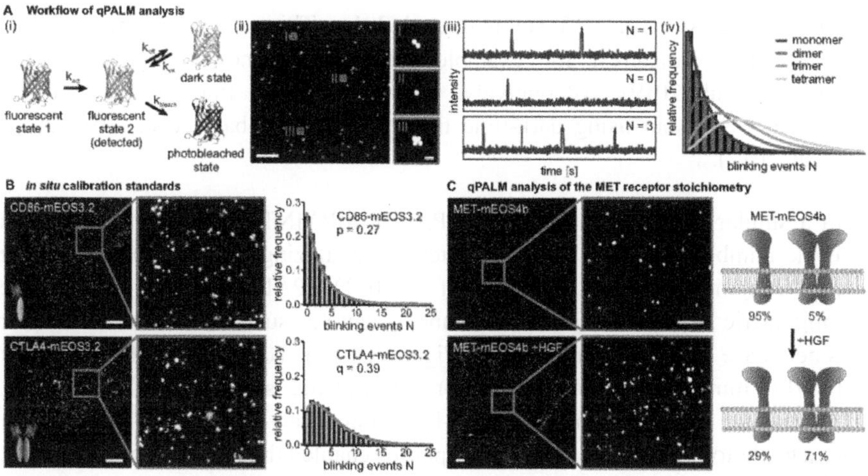

Figure 9.4. *Molecule counting with SMLM: quantitative PALM (qPALM) reports on membrane protein oligomerization. (A) Principle of qPALM. (i) Photoactivatable or photoconvertible fluorescent proteins reversibly switch between a fluorescent state and a dark state and finally photobleach. Here, qPALM with different variants of the photoconvertible fluorescent protein mEos is shown. The crystal structure in the scheme is mEos4b (PDB 6YLS). (ii,iii) After single-molecule localization, a super-resolved image is generated. Clusters that fulfil filtering criteria in size, shape and isolation are selected and the number of blinking events is extracted (a blinking event is defined as the recurrence of fluorescence). Scale bar 1 µm, insets 10 nm. (iv) Analyzing many clusters yields a frequency histogram that can be fitted with a probability function that reports on the oligomeric state of the molecule. (B) For robust stoichiometry analysis, calibration standards that report on the photobleaching probability p and the detection probability q are needed. Such calibration standards can be proteins of known stoichiometry (monomeric, dimeric) fused to the same fluorescent protein and expressed in cells. Scale bars 2 µm, insets 500 nm. Adapted from Baldering et al. (2019). Copyright CC BY-NC-SA 3.0. (C) qPALM analysis of the receptor tyrosine kinase MET in the absence and presence of activating ligand HGF. Cells treated with HGF show a significant increase in dimeric MET. Scale bars 2 µm, insets 1 µm. Adapted from Baldering et al. (2021); Copyright CC BY 4.0. For a color version of this figure, see www.iste.co.uk/sibarita/photonic.zip*

Quantitative PALM (qPALM) exploits that fluorescent proteins often show multiple emission events per single fluorophore (Avilov et al. 2014). In a multimeric complex, where each unit is labeled with a fluorescent protein, the total number of emission events can be related to molecule numbers using kinetic equations

(Figure 9.4) (Fricke et al. 2015; Hummer et al. 2016). This analysis can be calibrated with reference structures, for example, proteins of known oligomeric state (Baldering et al. 2019). Alternatively, intervals between fluorescence emission events can be cut out using dark-time thresholding (Annibale et al. 2010, 2011; Lee et al. 2012).

In a similar way, quantitative DNA-PAINT (qPAINT) obtains information on molecule numbers in single protein clusters by analyzing the fluorophore binding kinetics (Jungmann et al., 2016; Fischer et al., 2021). qPALM was used to determine the oligomeric state of various membrane receptors, such as the MET receptor tyrosine kinase (Baldering et al. 2021) (Figure 9.4(C)), Toll-like receptor 4 (Krüger et al. 2017), tumor-necrosis factor receptor 1 (Karathanasis et al. 2020; Weinelt et al. 2021), and the anion channel SLAH3 (Lehmann et al. 2021). In a very similar approach, photoswitchable organic fluorophores enabled the detection of monomeric and dimeric μ-opioid receptors (Möller et al. 2020). qPAINT was used to resolve the oligomeric state of the purinergic receptor Y2 (Joseph et al. 2021).

9.5. Conclusion

Single-molecule localization microscopy is beginning to revolutionize our understanding of cell biology. Next to providing images with a spatial resolution in the range of a few nanometers, a plethora of analytical approaches is available to extract quantitative information from SMLM data. Recent developments in imaging multiple targets are beginning to pave the route towards visualizing a large number of proteins in their functional context and thus establish SMLM for optical structural biology.

9.6. Acknowledgments

We thank Marius Glogger for providing super-resolution images of the endoplasmic reticulum. M.S.D. and M.H. acknowledge funding by the Deutsche Forschungsgemeinschaft (German Science Foundation; DFG; grants SFB 902, SFB 1177, HE 6166/17-1).

9.7. References

Albertazzi, L. and Heilemann, M. (2023). When weak is strong: A plea for low-affinity binders for optical microscopy. *Angewandte Chemie*, 62(35), e202303390.

Andronov, L. et al. (2016). ClusterViSu, a method for clustering of protein complexes by Voronoi tessellation in super-resolution microscopy. *Scientific Reports*, 6, 24084.

Ankerst, M., Breunig, M.M., Kriegel, H.-P., Sander, J. (1999). OPTICS: Ordering points to identify the clustering structure. In *Proceedings of the 1999 ACM SIGMOD International Conference on Management of Data*. Association for Computing Machinery, New York.

Annibale, P. et al. (2010). Photoactivatable fluorescent protein mEos2 displays repeated photoactivation after a long-lived dark state in the red photoconverted form. *The Journal of Physical Chemistry Letters*, 1(9), 1506–1510. doi: 10.1021/jz1003523.

Annibale, P. et al. (2011). Quantitative photo activated localization microscopy: Unraveling the effects of photoblinking. *PloS One*, 6(7), e22678.

Avilov, S. et al. (2014). In cellulo evaluation of phototransformation quantum yields in fluorescent proteins used as markers for single-molecule localization microscopy. *PloS One*, 9(6), e98362.

Baldering, T.N. et al. (2019). Synthetic and genetic dimers as quantification ruler for single-molecule counting with PALM. *Molecular Biology of the Cell*, 30(12), 1369–1376.

Baldering, T.N. et al. (2021). CRISPR/Cas12a-mediated labeling of MET receptor enables quantitative single-molecule imaging of endogenous protein organization and dynamics. *iScience*, 24(1), 101895.

Banterle, N. et al. (2013). Fourier ring correlation as a resolution criterion for super-resolution microscopy. *Journal of Structural Biology*, 183(3), 363–367.

Betzig, E. (1995). Proposed method for molecular optical imaging. *Optics Letters*, 237. doi: 10.1364/ol.20.000237.

Betzig, E. et al. (2006). Imaging intracellular fluorescent proteins at nanometer resolution. *Science*, 1642–1645. doi: 10.1126/science.1127344.

Campello, R.J.G.B. et al. (2013). Density-based clustering based on hierarchical density estimates. In *Pacific-Asia Conference on Knowledge Discovery and Data Mining*. Springer, 160–172.

Descloux, A. et al. (2019). Parameter-free image resolution estimation based on decorrelation analysis. *Nature Methods*, 16(9), 918–924.

Descloux, A. et al. (2020). Addendum: Parameter-free image resolution estimation based on decorrelation analysis. *Nature Methods*, 16(9), 1061–1063.

Descloux, A. et al. (2021). Parameter-free rendering of single-molecule localization microscopy data for parameter-free resolution estimation. *Communications Biology*, 4(1), 550.

Ehmann, N. et al. (2014). Quantitative super-resolution imaging of Bruchpilot distinguishes active zone states. *Nature Communications*, 5, 4650.

Endesfelder, U. et al. (2014). A simple method to estimate the average localization precision of a single-molecule localization microscopy experiment. *Histochemistry and Cell Biology*, 141(6), 629–638.

Ester, M. et al. (1996). A density-based algorithm for discovering clusters in large spatial databases with noise. In *Proceedings of 2nd International Conference on Knowledge Discovery and Data Mining (KDD-96)*. AAAI Press, 226–231.

Fischer, L.S. et al. (2021). Quantitative single-protein imaging reveals molecular complex formation of integrin, talin, and kindlin during cell adhesion. *Nature Communications*, 12(1), 919.

Fricke, F. et al. (2015). One, two or three? Probing the stoichiometry of membrane proteins by single-molecule localization microscopy. *Scientific Reports*, 5, 14072.

Fürstenberg, A. and Heilemann, M. (2013). Single-molecule localization microscopy-near-molecular spatial resolution in light microscopy with photoswitchable fluorophores. *Physical Chemistry Chemical Physics: PCCP*, 15(36), 14919–14930.

Greenfield, D. et al. (2009). Self-organization of the *Escherichia coli* chemotaxis network imaged with super-resolution light microscopy. *PLoS Biology*, 7(6), e1000137.

Gustafsson, M.G.L. et al. (1995). Sevenfold improvement of axial resolution in 3D wide-field microscopy using two objective lenses. In *Proceedings of SPIE*, 2412, 147–156. doi: 10.1117/12.205334.

Gustafsson, M.G.L. et al. (2000). Doubling the lateral resolution of wide-field fluorescence microscopy using structured illumination. In *Proceedings of SPIE*, 3919, 141–150. doi: 10.1117/12.384189.

Harwardt, M.-L.I.E. et al. (2020). Single-molecule super-resolution microscopy reveals heteromeric complexes of MET and EGFR upon ligand activation. *International Journal of Molecular Sciences*, 21(8). doi: 10.3390/ijms21082803.

Heilemann, M. et al. (2005). Carbocyanine dyes as efficient reversible single-molecule optical switch. *Journal of the American Chemical Society*, 3801–3806. doi: 10.1021/ja044686x.

Heilemann, M. et al. (2008). Subdiffraction-resolution fluorescence imaging with conventional fluorescent probes. *Angewandte Chemie*, 47(33), 6172–6176.

Hell, S.W. and Wichmann, J. (1994). Breaking the diffraction resolution limit by stimulated emission: Stimulated-emission-depletion fluorescence microscopy. *Optics Letters*, 19(11), 780–782.

Hess, S.T. et al. (2006). Ultra-high resolution imaging by fluorescence photoactivation localization microscopy. *Biophysical Journal*, 91(11), 4258–4272.

Hummer, G. et al. (2016). Model-independent counting of molecules in single-molecule localization microscopy. *Molecular Biology of the Cell*, 27(22), 3637–3644.

Joseph, M.D. et al. (2021). Quantitative super-resolution imaging for the analysis of GPCR oligomerization. *Biomolecules*, 11(10). doi: 10.3390/biom11101503.

Jradi, F.M. and Lavis, L.D. (2019). Chemistry of photosensitive fluorophores for single-molecule localization microscopy. *ACS Chemical Biology*, 14(6), 1077–1090.

Jungmann, R. et al. (2010). Single-molecule kinetics and super-resolution microscopy by fluorescence imaging of transient binding on DNA origami. *Nano Letters*, 10(11), 4756–4761.

Jungmann, R. et al. (2014). Multiplexed 3D cellular super-resolution imaging with DNA-PAINT and exchange-PAINT. *Nature Methods*, 11(3), 313–318.

Jungmann, R. et al. (2016). Quantitative super-resolution imaging with qPAINT. *Nature Methods*, 13(5), 439–442.

Karathanasis, C. et al. (2020). Single-molecule imaging reveals the oligomeric state of functional TNFα-induced plasma membrane TNFR1 clusters in cells. *Science Signaling*, 13(614). doi: 10.1126/scisignal.aax5647.

Kiuchi, T. et al. (2015). Multitarget super-resolution microscopy with high-density labeling by exchangeable probes. *Nature Methods*, 12(8), 743–746.

Klevanski, M. et al. (2020). Automated highly multiplexed super-resolution imaging of protein nano-architecture in cells and tissues. *Nature Communications*, 11(1), 1552.

Krüger, C.L. et al. (2017). Quantitative single-molecule imaging of TLR4 reveals ligand-specific receptor dimerization. *Science Signaling*, 10(503). doi: 10.1126/scisignal.aan1308.

Lee, S.-H. et al. (2012). Counting single photoactivatable fluorescent molecules by photoactivated localization microscopy (PALM). *Proceedings of the National Academy of Sciences*, 17436–17441. doi: 10.1073/pnas.1215175109.

Lehmann, J. et al. (2021). Acidosis-induced activation of anion channel SLAH3 in the flooding-related stress response of Arabidopsis. *Current Biology: CB*, 31(16), 3575–3585.e9.

Levet, F. et al. (2015). SR-Tesseler: A method to segment and quantify localization-based super-resolution microscopy data. *Nature Methods*, 12(11), 1065–1071.

Levet, F. et al. (2019). A tessellation-based colocalization analysis approach for single-molecule localization microscopy. *Nature Communications*, 10(1), 2379.

van de Linde, S. et al. (2008). Subdiffraction-resolution fluorescence imaging of proteins in the mitochondrial inner membrane with photoswitchable fluorophores. *Journal of Structural Biology*, 164(3), 250–254.

van de Linde, S. et al. (2010). The effect of photoswitching kinetics and labeling densities on super-resolution fluorescence imaging. *Journal of Biotechnology*, 149(4), 260–266.

Lutz, T. et al. (2018). Versatile multiplexed super-resolution imaging of nanostructures by quencher-exchange-paint. *Biophysical Journal*, 349a. doi: 10.1016/j.bpj.2017.11.1946.

Malkusch, S. et al. (2012). Coordinate-based colocalization analysis of single-molecule localization microscopy data. *Histochemistry and Cell Biology*, 137(1), 1–10.

Mihael, A. et al. (1999). OPTICS: Ordering points to identify the clustering structure. In *Proc. ACM SIGMOD'99 Int. Conf. on Management of Data*. ACM Press, 49–60.

Möller, J. et al. (2020). Single-molecule analysis reveals agonist-specific dimer formation of μ-opioid receptors. *Nature Chemical Biology*, 16(9), 946–954.

Moore, R.P. and Legant, W.R. (2018). Improving probes for super-resolution. *Nature Methods*, 659–660.

Mortensen, K.I. et al. (2010). Optimized localization analysis for single-molecule tracking and super-resolution microscopy. *Nature Methods*, 7(5), 377–381.

Mund, M. et al. (2018). Systematic nanoscale analysis of endocytosis links efficient vesicle formation to patterned actin nucleation. *Cell*, 174(4), 884–896.e17.

Narayanasamy, K.K. et al. (2021). Visualizing synaptic multi-protein patterns of neuronal tissue with DNA-assisted single-molecule localization microscopy. *Frontiers in Synaptic Neuroscience*, 13, 671288.

Narayanasamy, K.K. et al. (2022). Fast DNA-PAINT imaging using a deep neural network. *Nature Communications*, 13(1), 5047.

Nicovich, P.R. et al. (2017). Turning single-molecule localization microscopy into a quantitative bioanalytical tool. *Nature Protocols*, 12(3), 453–460.

Nieuwenhuizen, R.P.J. et al. (2013). Measuring image resolution in optical nanoscopy. *Nature Methods*, 10(6), 557–562.

Ovesný, M. et al. (2014). ThunderSTORM: A comprehensive ImageJ plug-in for PALM and STORM data analysis and super-resolution imaging. *Bioinformatics*, 30(16), 2389–2390.

Pageon, S.V. et al. (2016). Clus-DoC: A combined cluster detection and colocalization analysis for single-molecule localization microscopy data. *Molecular Biology of the Cell*, 27(22), 3627–3636.

Patterson, G.H. and Lippincott-Schwartz, J. (2002). A photoactivatable GFP for selective photolabeling of proteins and cells. *Science*, 297(5588), 1873–1877.

Pawley, J.B. and Masters, B.R. (2008). Handbook of biological confocal microscopy, third edition. *Journal of Biomedical Optics*, 13(2), 029902. doi: 10.1117/1.2911629.

Ripley, B.D. (1976). The second-order analysis of stationary point processes. *Journal of Applied Probability*, 255–266. doi: 10.1017/s0021900200094328.

Rust, M.J. et al. (2006). Sub-diffraction-limit imaging by stochastic optical reconstruction microscopy (STORM). *Nature Methods*, 3(10), 793–795.

Sage, D. et al. (2019). Super-resolution fight club: Assessment of 2D and 3D single-molecule localization microscopy software. *Nature Methods*, 16(5), 387–395.

Sahl, S.J. and Hell, S.W. (2019). High-resolution 3D light microscopy with STED and RESOLFT. In *High Resolution Imaging in Microscopy and Ophthalmology: New Frontiers in Biomedical Optics*, Bille, J.F. (ed.). Springer, Cham.

Sahl, S.J. et al. (2017). Fluorescence nanoscopy in cell biology. *Nature Reviews Molecular Cell Biology*, 685–701. doi: 10.1038/nrm.2017.71.

Sauer, M. and Heilemann, M. (2017). Single-molecule localization microscopy in eukaryotes. *Chemical Reviews*, 117(11), 7478–7509.

Schnitzbauer, J. et al. (2017). Super-resolution microscopy with DNA-PAINT. *Nature Protocols*, 12(6), 1198–1228.

Schröder, M.S. et al. (2021). Imaging the fibroblast growth factor receptor network on the plasma membrane with DNA-assisted single-molecule super-resolution microscopy. *Methods*, 193, 38–45.

Schueder, F. et al. (2024). Unraveling cellular complexity with transient adapters in highly multiplexed super-resolution imaging. *Cell*, 187(7), 1769–1784.e18.

Sengupta, P. and Lippincott-Schwartz, J. (2012). Quantitative analysis of photoactivated localization microscopy (PALM) datasets using pair-correlation analysis. *BioEssays: News and Reviews in Molecular, Cellular and Developmental Biology*, 34(5), 396–405.

Sengupta, P. et al. (2011). Probing protein heterogeneity in the plasma membrane using PALM and pair correlation analysis. *Nature Methods*, 8(11), 969–975.

Shannon, C.E. (1984). Editorial note on "Communication in the presence of noise". *Proceedings of the IEEE*, 1713–1713. doi: 10.1109/proc.1984.13079.

Sharonov, A. and Hochstrasser, R.M. (2006). Wide-field subdiffraction imaging by accumulated binding of diffusing probes. *Proceedings of the National Academy of Sciences of the United States of America*, 103(50), 18911–18916.

Spahn, C.K. et al. (2018). A toolbox for multiplexed super-resolution imaging of the *E. coli* nucleoid and membrane using novel PAINT labels. *Scientific Reports*, 8(1), 14768.

Steinhauer, C. et al. (2011). Nanoscopy using localization and temporal separation of fluorescence from single molecules. In *Biophotonics: Spectroscopy, Imaging, Sensing, and Manipulation*, Di Bartolo, B. and Collins, J. (eds). Springer, Dordrecht, 87–106. doi: 10.1007/978-90-481-9977-8_5.

Szymborska, A. et al. (2013). Nuclear pore scaffold structure analyzed by super-resolution microscopy and particle averaging. *Science*, 341(6146), 655–658.

Unterauer, E.M. et al. (2024). Spatial proteomics in neurons at single-protein resolution. *Cell*, 187(7), 1785–1800.e16.

Wardhan, R. and Mudgal, P. (eds) (2017). Membrane receptors and signal transduction pathway. In *Textbook of Membrane Biology*. Springer, Singapore, 293–344. doi: 10.1007/978-981-10-7101-0_9.

Weinelt, N. et al. (2021). Quantitative single-molecule imaging of TNFR1 reveals zafirlukast as antagonist of TNFR1 clustering and TNFα-induced NF-κB signaling. *Journal of Leukocyte Biology*, 109(2), 363–371.

Werbin, J.L. et al. (2017). Multiplexed exchange-PAINT imaging reveals ligand-dependent EGFR and Met interactions in the plasma membrane. *Scientific Reports*, 7(1), 12150.

Wolter, S. et al. (2010). Real-time computation of subdiffraction-resolution fluorescence images. *Journal of Microscopy*, 237(1), 12–22.

Wu, Y. and Shroff, H. (2018). Faster, sharper, and deeper: Structured illumination microscopy for biological imaging. *Nature Methods*, 15(12), 1011–1019.

Yi, J. et al. (2016). madSTORM: A superresolution technique for large-scale multiplexing at single-molecule accuracy. *Molecular Biology of the Cell*, 27(22), 3591–3600.

10

Image Processing and Image Analysis in Microscopy

Daniel SAGE[1] and Anaïs BADOUAL[2]

[1] *Biomedical Imaging Group and Center for Imaging, École Polytechnique Fédérale de Lausanne (EPFL), Switzerland*

[2] *SAIRPICO, Centre INRIA de l'Université de Rennes, France*

10.1. Introduction

In recent years, microscopy images have played a crucial part in the life sciences. They are the privileged witnesses for the observation at the molecular level of the great many and very complex intra-cellular interactions. As a corollary, the quantitative analysis of these images has become essential to improve the understanding of this cellular machinery.

Throughout this book, the reader has discovered a very wide range of microscopy techniques, thus demonstrating the ingenuity of scientists striving to challenge the knowledge of the living world ever further. While the acquisition of microscopic information varies greatly from one microscopy modality to another, they all now share a common denominator: the digital world. Indeed, since the 2000s, behind every microscope there are now automated systems, digital sensors, digital cameras and of course one or more computers. The latter are present throughout the entire process from image acquisition to the interpretation of their content. The nature of experiments has therefore evolved from purely qualitative observations to quantitative analyses on computers.

For a color version of all figures in this chapter, see www.iste.co.uk/sibarita/photonic.zip.

Photonic Imaging for Biology,
coordinated by Jean-Baptiste SIBARITA. © ISTE Ltd 2025.

The automatic quantification of images and the extraction of relevant information have become a major challenge for the biosciences. Discoveries in biology, progress in the knowledge of the cell, are often significantly linked to the analysis of microscopy images. This is why the demand for dedicated computer tools is increasingly important. However, image analysis solutions that are versatile enough and capable of absorbing the huge flow of data (Ouyang and Zimmer 2017) still remain very uncommon, especially with the new modalities such as super-resolution microscopy or large field of view imaging.

The challenges that must be addressed are at the same time technological, logistical, as well as methodological with the consequent developments that have been gathered and centered around a new discipline called *computational bioimaging* or *bioimage informatics* (Myers 2012; Meijering et al. 2016). This discipline is boosted by the synergies and transversal exchanges between the fields of biology, microscopy, signal processing and computer vision, combining approaches known as *model-based* and, more recently, approaches referred to as *data-driven*[1] (Meijering 2020).

In this chapter, we address the main aspects of computational bioimaging applied to the microscopy of living organisms. After setting the current context of microscopy imaging (section 10.2), we will present the most notable operations of image processing (section 10.3) and image analysis (section 10.4) of microscopy under the dual umbrella of model-based and data-driven approaches. The goal is to give the reader the general principles underlying algorithms and the terms used in the field allowing for more in-depth research. We will conclude with a perspective on the challenges of microscopy image processing and analysis (section 10.5).

10.2. Context

10.2.1. *Multidimensional images*

Microscopy images have nothing in common with images or photos that we usually come across in our daily lives, both in terms of size and content. They are rather regular groupings of measurements that reflect physical or biological phenomena, and which are intended to be quantified by a computer. In bioimaging, an image is an object that can have up to five dimensions (5D). The data originating from the lateral plane form a 2D image with two XY axes, to which is added the spatial dimension Z, obtained by scanning in the optical axis. The whole constitutes a volume

1. In this chapter, we use the term *model-based* for approaches designed by engineers based on physical, mathematical or biological modeling. These *model-based* methods are commonly referred to as conventional algorithmic methods as opposed to the newer machine learning techniques that we call *data-driven* methods. These machine learning techniques, including deep learning, are exclusively built around data without an explicit model.

of 3D data that enables the three-dimensional nature of the specimens to be captured. It should be noted that the axial resolution (Z) is often worse than the lateral resolution (XY) due to the technical limitations of microscopes. The quantification of dynamic processes requires the acquisition of time-lapse volumes at regular time intervals, thus forming a 4D space–time object. Finally, since the 2000s, fluorescence has become widely used in biology because it enables specific cellular structures to be isolated in different color channels of the image using markers of different wavelengths. This is the fifth dimension. This dimension is sometimes also derived from spectral imaging techniques (Zimmermann et al. 2003). For a specific experiment, it is therefore not uncommon to be dealing with 5D images (see Figure 10.1).

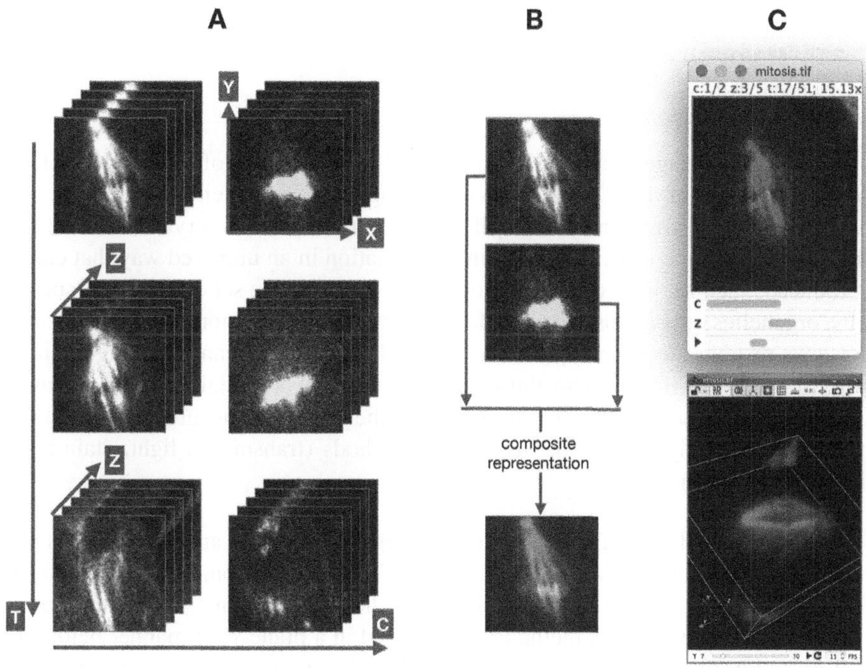

Figure 10.1. *Views of a 5D image. A. Illustration of a 5D image with five focal planes (Z), three time frames (T) and two fluorescent channels (C). B. Two 2D fluorescent channels (XY) of the 5D image are combined into a "composite" image with red and green color mapping. C. Top: ImageJ viewer of a 5D image with three sliders (C, Z, T) (Schindelin et al. 2015). Bottom: Icy viewer of a 5D image with volumetric rendering and a slider (T) (de Chaumont et al. 2012). Source: "mitosis.tif" 5D image provided as an example with the ImageJ software*

An image f is mathematically modeled by a function of several variables x, y, z, t, c which, in the numerical world, take their values in a finite space (see equation [10.1]). In the rest of this chapter, we use vector notation where $\mathbf{x} =$

$\{x, y, z, t, c\}$. The intensity $f[\mathbf{x}]$ at a given position $\mathbf{x}$ is discretized over a number of gray levels (8 bits in general, 16 bits in microscopy (see Figure 10.3(A), quantization)):

$$\begin{aligned} f : \Omega^5 &\rightarrow \mathcal{D} \\ \mathbf{x} &\rightarrow f[\mathbf{x}] \end{aligned} \qquad [10.1]$$

This mathematical formalism is used for the design of more efficient and generalizable algorithms, such as for the computation of the gradient vector (multidimensional derivative) which is the basis of edge detection algorithms.

10.2.2. *Major challenges in bioimaging*

Scientific challenges

Quantitative image analysis in biology has a wide range of applications, from simple counting of cell nuclei, to measuring protein–protein interactions, not forgetting the construction of the cell line of a *C elegans* embryo. The scientific challenge therefore consists of extracting information in an unbiased way that can be reproduced, such as, for example, for detecting a very dense set of particles (such as cells, organelles or proteins) and also for characterizing their morphology, dynamics, divisions, response to gene expression, or still the localization and tracking of single molecules, all from images with almost indistinguishable or partially degraded signals. An additional issue inherent to the study of the living is the great variability of the objects of study and of the acquisition methods (transmitted light, staining of histological sections, etc.).

By accounting for all dimensions of the image (5D), image analysis presents new methodological challenges. For example, for 3D segmentation, it is more reliable to extract a 3D-modeled object directly from the entire volume, rather than using traditional 2D segmentation methods that proceed in a plane-wise manner before the final assembly to obtain a 3D object. Naturally, the analysis with all dimensions is more demanding in computational terms.

Technical challenges

Modern microscopes have the ability to generate multidimensional images with high spatiotemporal resolution. The resulting multidimensional images are very large, sometimes of several terabytes (TB). As an example, light-sheet microscopes can produce several TB/h of data. In addition to the scientific challenges mentioned above, this massive data, especially 4D and 5D, complicates all operations, even the simplest ones such as archiving or transfer, especially for image processing and analysis operations.

A significant aspect in bioimaging is the visualization of images and data obtained by the analysis. Nonetheless, the visualization of 3D, 4D and 5D data is not easy on the usual 2D media, such as screens or paper. Even 3D volumetric rendering techniques only partially reflect the complex three-dimensional reality of the biological objects observed.

Having reliable annotated data is crucial to quantitatively evaluate the image analysis methods developed by the community and to train those based on deep learning (i.e. data-driven). The major issue is that manually annotating 3D, 4D and 5D images is an extremely tedious or even impossible task, mainly due to the obstacles to their visualization. Consequently, it cannot be reliably performed and exhibits significant intra- and inter-experimentalist variabilities. For a vast majority of applications on living cells, there is a recurring lack of verified annotations that could be used as a basis for learning. To overcome this challenge, ingenious image processing solutions are needed, such as the use of techniques to reduce the dimensionality of data (principal component analyses, kymographs or auto-encoders), the use of new visualization methods such as virtual reality, or even the development of large annotated synthetic databases.

The analysis of cellular processes therefore requires advanced methods for the manipulation of terabytes of data and solving these scientific challenges. Furthermore, as technology evolves in microscopy (microscopes, fluorescence probes (Berlin et al. 2015)), images are being enriched with new information that requires continuous developments in image processing and analysis.

10.2.3. *Software for bioimaging*

A lot of software is open source in bioimaging (Lucas et al. 2021). This has been produced by and for the bioimaging research community that has gathered around a very active forum image.sc[2]. This collaborative work makes it possible to follow technological advances in microscopy as closely as possible and to meet the precise needs of biologists with the development of ready-to-use tools.

In this context, ImageJ is a software program with an exclusive story[3]. Its creator, Wayne Rasband of the National Health Institute, developed it over a period of 20 years. ImageJ was quickly adopted to become almost universal in bioimaging. Due to its open architecture and code written in a very popular language (Java), hundreds of developers from all over the world have been able to add scripts, macros and plugins to satisfy most of the needs of microscopy image analysis. ImageJ is the source of a

2. image.sc: https://forum.image.sc.
3. ImageJ: https://imagej.nih.gov/ij/.

software ecosystem (Schindelin et al. 2015) centered around the new ImageJ2[4]. There are several distributions of ImageJ, including the very popular Fiji[5] which is today installed on most biologist computers.

Other open-source projects for a generalist platform for bioimaging have been initiated:

– Icy[6] from the Institut Pasteur, coded in Java, is capable of visualizations and native operations in 5D.

– QuPath[7] by Pete Blankhead is very popular for examining digital pathology slides.

– CellProfiler[8] from the Broad Institute provides image analysis pipelines for biologists.

– Ilastik[9], mainly developed at the University of Heidelberg, includes a comprehensive set of machine learning algorithms.

– Napari[10], initiated by the Chan-Zuckerberg Biohub, is a multidimensional image analysis software written in Python.

While Java has long been the language of choice for bioimaging, Python is now often preferred mainly because of the attractiveness of Jupyter Notebooks, as well as because of the native access to powerful deep learning frameworks such as TensorFlow[11] and PyTorch[12]. The new bioimaging platform, Napari, is a driving force of this evolution towards Python, by combining an efficient interactive multi-dimensional viewer and a large number of plugins for image analysis.

Although there is a significant offer of software programs, it only partially covers the specific needs of image analysis and it is often necessary to use specific software developments for a given type of image and for a given application. Within this context, it is essential that scientists adopt the proper practices of open science, by sharing data (Ellenberg et al. 2018), analysis protocols, software source codes

4. ImageJ2: https://imagej.net/software/imagej2/.
5. Fiji: https://imagej.net/software/fiji/.
6. Icy: http://icy.bioimageanalysis.org/.
7. QuPath: https://qupath.github.io/.
8. CellProfiler: https://cellprofiler.org/.
9. Ilastik: https://www.ilastik.org/.
10. Napari: https://napari.org/.
11. TensorFlow: https://www.tensorflow.org/.
12. PyTorch: https://pytorch.org/.

(Carpenter et al. 2012), deep learning models[13] and validation methods in the spirit of collaborative research (Cohen 2014).

10.2.4. *Image analysis pipeline*

In computer vision, an image analysis pipeline generally includes two phases (see Figure 10.2): image processing operations (section 10.3) and image analysis operations (section 10.4). These two phases are found in microscopy, to which are sometimes added an image reconstruction step, which consists of numerically combining a set of measurements or images.

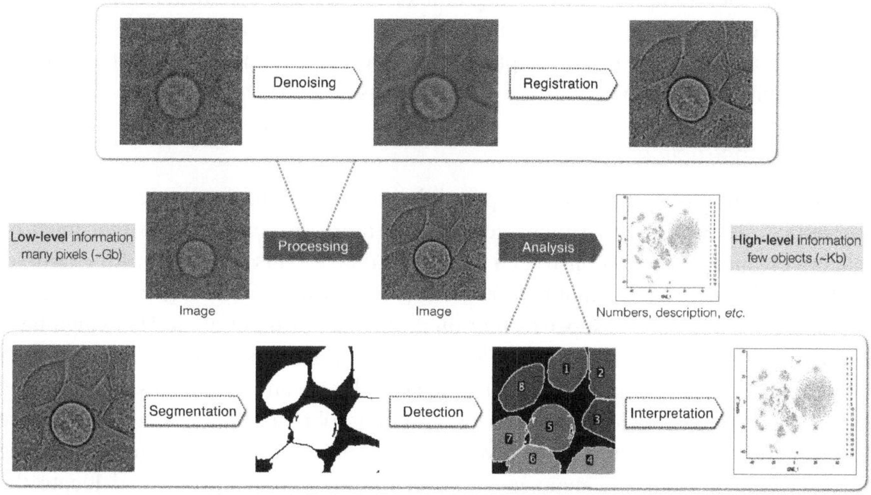

Figure 10.2. *An example of an image analysis pipeline. A pipeline is often made up of different steps that allow shifting from very low-level information (with a very large amount of pixel values) to very high-level information (with a very limited number of results). Only the key steps that are generally common to all image analysis pipelines are depicted in this figure. Depending on the application, other steps may be required such as image reconstruction, image registration, object classification or object tracking. Source: image from the Cell Tracking Challenge, http://celltrackingchallenge.net*

In order to maximize the performance of the image analysis pipeline for a given application, image acquisition must be integrated from the design stage of the project by seeking the best compromise of the acquisition parameters. This avoids, for example, having too high a spatial resolution and insufficient time resolution if the goal is to follow fast movements.

13. Biomage Model Zoo: https://bioimage.io.

In general, an image analysis pipeline is broken down into basic steps where each step can be performed with tools ready for use. On the other hand, it is therefore imperative to validate the pipeline in its entirety to limit the inexorable propagation of errors in the pipeline and reduce the dependency of the parameters of each step.

Image processing and image analysis

Image processing operations are digital transformations of images that take an image on input and output an image, generally of the same dimension and size, and whose purpose is to facilitate the extraction of information. These are thus often pre-processing operations whose objective is to improve the structures of interest in the images, for example, by removing blur or reducing noise, while improving the structures of interest. Strictly speaking, image analysis operations are those that have an input image and non-image output results, such as a number of detected objects, statistical information about the size of the objects or their dynamics, etc.

Image reconstruction

For a number of recent microscopy modalities, access to the image is not direct. The optical detectors of the microscope record a set of measurements that must be collectively analyzed to reconstruct a computed image. This is the case, for example, of the SIM (Structured Illumination Microscopy) and SMLM (Single-Molecule Localization Microscopy) super-resolution modalities (Valli et al. 2021), which have a computer reconstruction phase that leads to overpassing the physical limits of light diffraction. In SIM (Gustafsson 2000), nine spatially modulated images are generally acquired and then merged in the Fourier domain to obtain a single image with twice the resolution (see Chapter 7). In SMLM, thousands of images of molecules made fluorescent in a stochastic manner are recorded. Algorithms detect and locate the positions of the sources very accurately to reconstruct a map of the centers that forms an image about 10 times better resolved (see Chapter 9). Image reconstruction algorithms play a key role in the quality of the reconstructed image (Sage et al. 2019).

The role of algorithms is concomitant with any reconstruction process. Once the physical model is properly characterized, the reconstruction is mathematically formalized as the solution of a matrix equation called *inverse problem*, in which the unknowns are the values of the pixels to be reconstructed. This inverse problem is often ill-posed because the information is partial, the model is approximate and the level of noise is often high.

Traditional model-based approaches versus recent data-driven approaches

The different steps of image reconstruction, processing and analysis can be performed separately or all together using *model-based* or *data-driven* approaches.

Model-based approaches (deconvolution, active contours, Markov Random Fields), recognized as traditional approaches, are based on physical or biological models. Their main advantage is to easily integrate prior knowledge of the process under study (shape, trajectory, possible distortion, kinetics), and due to their mathematical formalism, they allow for error analysis and an interpretation of the results obtained. The challenge with these approaches is to succeed in correctly modeling the underlying processes with appropriate levels of detail and realism. This requires some *ingenuity* from the designer.

The more recent data-driven approaches (such as for instance machine learning and deep learning) for their part are based on *learning* statistical models by accumulating information from a very large number of example data. These methods adapt better to the complexity of data and therefore generally perform better than model-based approaches. Once trained, they are also more computationally efficient. On the other hand, they pose a challenge with regard to the aspects of generalizability, explainability and interpretability (Meijering 2020) and the very high computational cost of their learning phase.

The choice of one or the other of these two approaches depends on the complexity of the task to be accomplished, the degree of variability of data, the prior knowledge of the underlying physical models, data availability and the expected interpretability of the results. One idea that is currently gaining momentum is to combine these two paradigms to obtain the best of both worlds (*hybrid* methods) (section 10.5).

10.3. Biomage processing

Digital image processing is a well-documented field (Russ 2002), and many image processing software programs include a wide range of tools. Many of the tools available are limited to 2D images, although many processings would benefit from being applied to other dimensions. In this section, we first present the fundamental operations of image processing, and we then develop three common applications for microscopy.

10.3.1. *Core processings*

Pixel-wise image processing operations are the simplest. They proceed by transforming the image at the level of the individual pixel by changing their value independently of the neighboring pixels. These operations are primarily useful for human image observation, for example, to correct photobleaching or to improve contrast by adjusting the histogram or by pseudo-color (see Figure 10.3A). More complex, filtering transformations are widely used in image processing and analysis because they make it possible to determine the value of each pixel of the output image based on a block of pixels in the input image. The neighbors of the pixel in question are generally the ones that will inform about the local context.

The common image processing filters include the family of linear filters that are entirely characterized by a mathematical operation: the convolution of two functions, f being the input image and h being the kernel of the filter, called the convolution kernel. The result of a convolution is a function g which is an image of the same size as f. The support Ω of the filter kernel h is critical to take more or less neighborhood context. The smallest 2D-centered support is 3 x 3 (see Figure 10.3B). The combination of convolutional operators into very large networks with multiple layers form convolutional neural networks (CNNs) that are at the heart of deep learning methods (Goodfellow et al. 2016). The discrete convolution is written as:

$$\begin{aligned} g[\mathbf{x}] &= f[\mathbf{x}] * h[\mathbf{x}] \\ &= \sum_{\mathbf{k}\in\Omega} f[\mathbf{k}]\, h[\mathbf{x}-\mathbf{k}]. \end{aligned} \quad [10.2]$$

The values or coefficients of the filter kernel h determine the effect of the filter. For instance, with given coefficients, we can make the image sharper with better defined edges, which will consequently amplify the noise. By choosing other coefficients for the filter kernel h, an integrator/averaging effect can be obtained such that it leads to a reduction in noise but will also make the image more blurry (see Figure 10.3B). These filters include the Gaussian filter which is the most widely used because it allows doing just about anything by the setting of a single parameter σ (standard deviation of the Gaussian) that gives access to the different scales and structures of different sizes in an image (see Figure 10.3C).

Linear filters (convolutions) with large supports can be efficiently performed in a dual domain to the spatial domain of the images: this is the frequency domain called the Fourier domain. It is common practice in signal processing to switch to a frequency representation of the signal using the Fourier transform, even when addressing multidimensional signals such as images (Broughton and Bryan 2018). The fast Fourier transform (FFT) algorithm ensures the fast and loss-free conversion of a domain. As an example, all convolution operations in deconvolution (section 10.3.3) are performed in the frequency domain to be faster.

Mathematical morphology filters, called rank filters, are another class of (nonlinear) filters that modify the shape of objects in the image with a neighborhood $\mathbf{W}$ known as a structuring element (Soille 2013). In this context, the two basic operations are the minimum filter and the maximum filter, also called erosion and dilation, which can be chained together to simplify images. Rank filters are used a lot as segmentation pre-processing, for example, to fill holes in an object or to remove a non-uniform background.

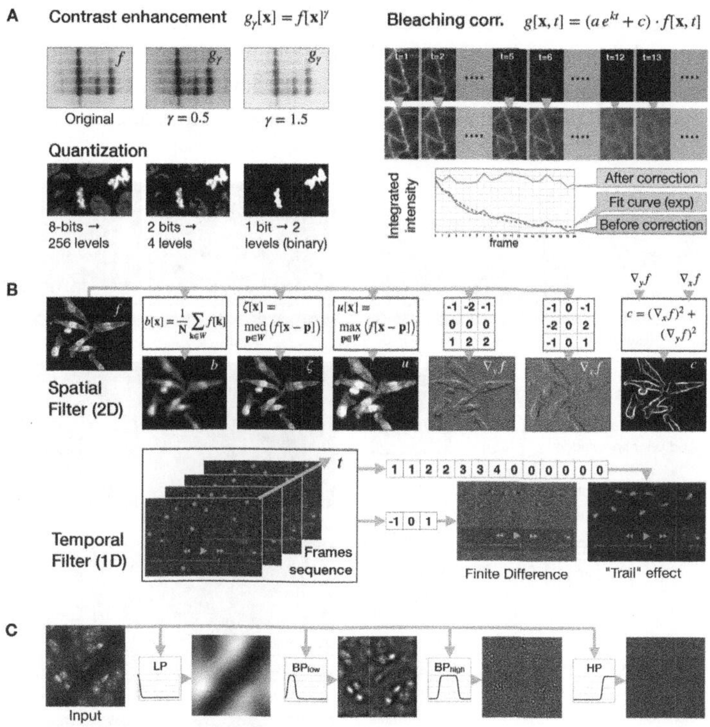

Figure 10.3. *Core operations in image processing. A. Pixel-wise operations. Left: contrast improvement in an image by using a Gamma correction function and simplifying an 8-bit image by quantization on a reduced number of bits. Right: correction of the loss of the fluorescent signal due to photobleaching with a decreasing exponential function parameterized by* a, c *and* k *(red dotted line); time series of images before correction in blue and after correction in green. B. Top: 2D spatial filters. From left to right. Blurred image* b *resulting from a mean filter (box filter) of size* $N \times N$ $(N = 7)$*. Images* ζ *and* u *obtained by applying the median and maximum nonlinear filters, respectively, to a neighborhood* $\mathbf{W}$*. The median filter is very helpful for denoising and for reducing details in the image without losing edges. Edge detection using the Sobel filter* c *by summing images* ∇_y *and* ∇_x *that were obtained by 3x3 finite-difference linear filters. Bottom: example of 1D linear filters in time, applied to a sequence of images, a filter with a "trail" effect, and a finite difference filter. C. Image decomposition with, from left to right, a lowpass filter (LP) to find a non-flat background; a bandpass filter set to low* (BP_{low}) *to have the nuclei; a bandpass filter set to high* (BP_{high}) *to have the small particles; a highpass filter (HP) to show the noise*

10.3.2. *Denoising of microscopy images*

Due to physical and technical limitations, the acquisition of a microscopy image is always the result of a compromise between spatial and/or temporal resolution, SNR and reduction of phototoxicity. Thereby, the low illumination to limit photobleaching and protect the living sample induces a high noise level with respect to the signal. Therefore, numerical denoising algorithms will be applied to reduce noise.

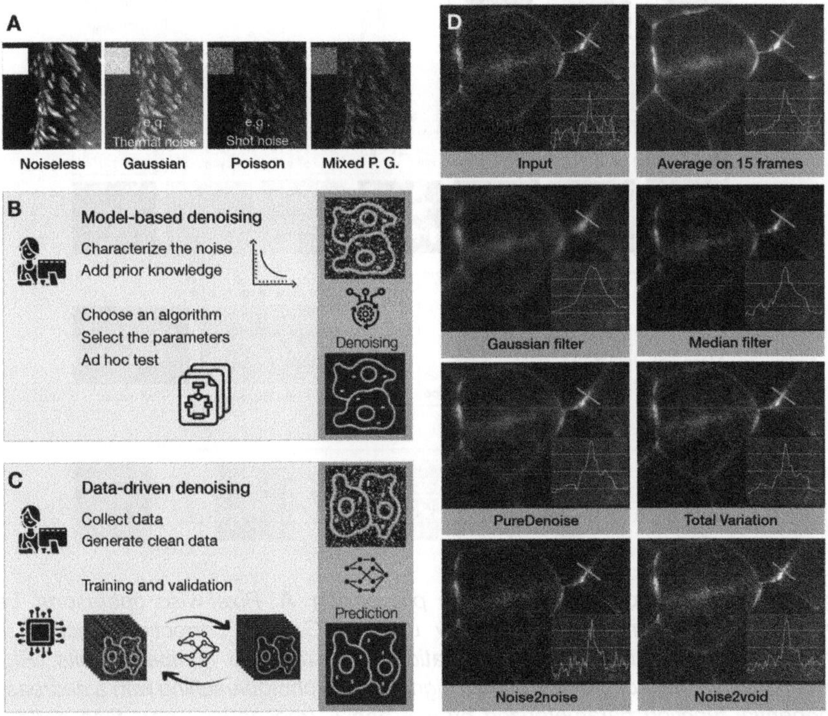

Figure 10.4. *Denoising of microscopy images. A. Examples of different types of noise in fluorescence microscopy: Gaussian, Poisson, a mixture of Gaussian and Poisson. Source: Cell Image Library CIL 38985. B. Design of a model-based denoising algorithm where the developer's involvement is key to best characterizing the noise model. C. Design of a data-driven algorithm where the prediction is based on a neural network trained with a large number of examples. D. Examples of denoising results with several algorithms applied to the "input" image. The denoising result at the top right was obtained by acquiring 15 noisy images of the same (fixed) structure and averaging them*

For practical reasons, we present here some denoising techniques by limiting ourselves to the 2D case. Nevertheless, the algorithms should be applied together with the other dimensions in order to be effective at better exploiting contextual information.

Source of noise

In microscopy, there are many sources of intrinsic noise during acquisition. A dominant source of noise in fluorescence microscopy is shot noise, which originates from the process of counting photons. It is stochastically modeled by a Poisson distribution. There are other sources of noise created by the electronics for reading the signal, such as dark current or readout noise which follow a Gaussian distribution (see Figure 10.4A). In real conditions, the Poisson/Gaussian distribution of noise is poorly known, which makes it difficult to choose and parameterize denoising algorithms.

Model-based methods

Denoising is based on two assumptions: 1) the noise is random and 2) the noise is independent from one pixel to another.

Since noise mostly occurs in high frequencies, a lowpass filter such as the *Gaussian* filter can be applied as a first approach. This linear filter reduces the high signal frequencies, but at the same time, it smooths out the details and contours of objects. It acts a bit as if the resolution of the image were reduced, which is sometimes very dearly earned during the acquisition phase. This major disadvantage of the Gaussian filter explains the development of more selective filtering methods that tend to better preserve the edges, such as the very basic *median* filter (see Figure 10.3B), the iterative anisotropic diffusion filter, the *mean-shift* aggregation filter, the bilateral filter and the inverse filter with total variation regularization (TV) (Meiniel et al. 2018). Filtering in a domain suitable for images is a relevant alternative, for example, in the *wavelet* domain which has excellent signal decorrelation properties with respect to noise (Luisier et al. 2010). Pyramidal image decomposition in wavelet bases is widely used in biomedical image processing because it provides efficient access to multi-scale analysis. In bioimaging, the *non-local mean* algorithm (Boulanger et al. 2009) is another approach that uses the redundancies of local structures in the data (self-similarities), which are very common in biology where images are often composed of a juxtaposition of fairly similar patches.

With all these so-called model-based methods (see Figure 10.4B), parameterization can prove difficult without an explicit characterization of the noise sources and structures of interest. This is why machine learning approaches solely based on data themselves have become very popular because they require no mathematical modeling of noise (Laine et al. 2021).

Data-driven methods

The machine learning approach is based on a training phase (see Figure 10.4C) which is an optimization step of the parameters of the function transforming the noisy image into a denoised image. In deep learning, this transformation function is built with an artificial neural network comprising millions of parameters to be optimized. Once the network has been trained, a prediction has simply to be applied to denoise a new image acquired under the same experimental conditions. Deep learning methods are currently the most efficient of the denoising methods (Kefer et al. 2021) (see Figure 10.4D).

In the supervised mode, training requires a large number of pairs of images: a noisy image and a noise-free image. Since noise-free images are not available in real conditions, other means must be found to obtain these pairs of images, either by simulation by adding noise to low-noise images, or by acquiring two images simultaneously, one with low illumination and one with high illumination, as in the CARE method (Weigert et al. 2018).

More recently, several self-supervised methods have been proposed for efficient denoising without clean noise-free images. Among these methods, the pioneer is Noise2Noise, developed by Lehtinen et al. (2018), which uses pairs of images for training, each consisting of two independent realizations. Noise2Void (Krull et al. 2019) and Noise2Self (Batson and Royer 2019) operate with another strategy to form pairs of images with the noisy image and the noisy image altered by randomly deleting one pixel. These methods can therefore very efficiently predict the erased pixel from the neighborhood. They have given rise to many variations: Noise2Same, Noise2Score, Noise2inverse, StructN2V, etc.

10.3.3. *Deconvolution of microscopy images*

Image formation model and PSF

Image formation in fluorescence microscopy follows the physical laws of light propagation that can be mathematically modeled by a convolution between the emitted light signal and the point-spread function (PSF) to which noise is added (see Figure 10.5A). The PSF represents the response of a microscope to a point light source.

Therefore, in 3D, the PSF integrates the deformations of each focal plane which gives it this typical double cone (hourglass) shape (see Figure 10.5A). The size of the PSF determines the spatial resolution of at best 250 nm laterally and 900 nm axially, and it determines the ability of an optical system to separate (resolve) two point light sources. Accurate knowledge of the PSF is essential for image reconstruction. A PSF can be either obtained experimentally by the acquisition of a z-stack of fluorescent microbeads, or analytically by numerical simulation from the optical parameters of the microscope (Kirshner et al. 2013).

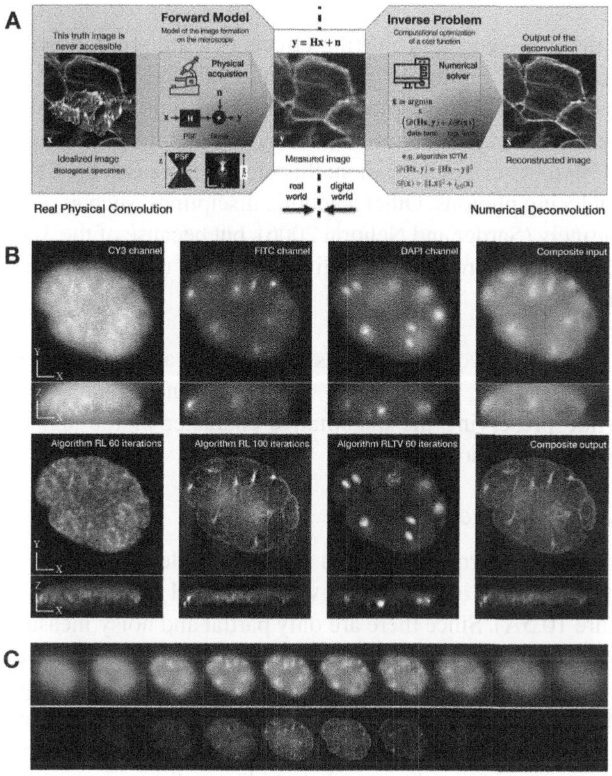

Figure 10.5. *Deconvolution of microscopy images. A. Formalization of deconvolution as an inverse problem. The forward model models the physical acquisition of the image by the microscope. The problem thus defined is solved by a numerical optimization method. The ICTM algorithm given as an example optimizes a criterion in the sense of the least squares with a Tikhonov regularization and a non-negativity constraint. B. 3D deconvolution of three fluorescent channels, CY3, FITC, DAPI. The sample is a C elegans embryo in which the first eight cell nuclei can be identified in DAPI (blue). The color results are a composition of the three fluorescent channels in a color composite image in red, green and blue. The images are maximum intensity projections along XY and XZ. Top: source image; bottom: image after deconvolution with an RL or RLTV algorithm. C. 3D deconvolution of the same sample. Presentation of a series of images in* Z *in the composite color mode showing the effect of reducing out-of-focus blurring. Top: images before deconvolution; bottom: images after deconvolution. All deconvolution images were generated with the open-source software DeconvolutionLab2 (Sage et al. 2017)*

Outcome of the deconvolution

The shape of the PSF generates blurring effects that are much more pronounced in the axial direction (see Figure 10.5A); the further away from the focal plane, the more the structures are drowned in an out-of-focus blur. Deconvolution results in reversing the effect of physical convolution by attempting to reassign the signal dispersed by the PSF to its original location; this greatly increases the SNR and signal dynamics, without losing photons. Other scientific disciplines, such as astronomy, use deconvolution routinely (Sarder and Nehorai 2006), but because of the 3D processing, this is the application in microscopy that gives the most remarkable deconvolution results (see Figure 10.5C).

At the same time, deconvolution acts as a denoising and contrast enhancement method, and primarily it considerably reduces out-of-focus blurring (see Figure 10.5B) while preserving the quantitative character of the signal. This thus slightly improves the axial resolution.

Deconvolution as an inverse problem

Deconvolution can be formalized as an inverse problem in which the forward model is a convolution of the observation $\mathbf{x}$ with a filter $\mathbf{H}$ (PSF) to which a noise $\mathbf{n}$ is added (see Figure 10.5A). Since there are only partial and noisy measurements, the problem is ill-posed and can lead to a non-unique solution. Regularizations should be added to restrict the solution to a space that is a priori acceptable. Two regularizations are common, Tikhonov regularization for rather smooth signals, and total variation (TV) regularization for signals composed of piecewise functions. For model-based methods, there are two existing classes of algorithms: direct methods (regularized inverse filter) which are fast but allow only little control over regularization, and iterative methods that gradually add content while adapting the regularization effect, iteration after iteration. In the iterative algorithms, the reference algorithm is the Richardson–Lucy (RL) deconvolution method (Richardson 1972). Many other algorithms can be found such as the Richardson–Lucy total variation (RLTV) (Dey et al. 2006) or iterative constraint Tikhonov–Miller (ICTM) (Van Kempen (2000)) used for the figure (see Figure 10.5C).

Deconvolution is a great example of how the computer can work jointly with the microscope to produce higher-resolution images with a better SNR by means of a process of digital restoration. Deconvolution has limitations that can generate reconstruction artifacts when the noise is too high or when the linear model of the PSF poorly approximates the nonlinear optical system. Deconvolution remains a rather expensive process in terms of computational time. Deconvolution is still in development to develop faster algorithms that can work with nonlinear models. The other direction of development consists of using deep learning, which for the moment faces a lack of properly resolved ground truth data.

10.3.4. *Image registration*

Image registration algorithms alter the coordinates of one or more destination images to align with a source image, either in 2D or 3D. In microscopy, the source and destination data can come from different detectors, different modalities, different time points or sometimes even from alignment with atlases.

Image registration is very commonly needed in bioimaging to be able to geometrically align inconsistent data. CLEM correlative imaging is a good example of this, by matching high-resolution electron microscopy images to light microscopy images (Anderson et al. 2019). This is a task that can be particularly complex when a transformation has to be established between two modalities whose resolution and content are different.

The prior knowledge of a transformation model is an essential element for registration. A distinction is made between rigid models, which correspond to a global affine transformation, and more complex elastic models, which correspond to local deformations.

There are two families of algorithms in image processing, namely intensity-based and feature-based algorithms. Intensity-based algorithms examine the pixel-to-pixel relationships of images to compute the affine transformation matrix or to estimate local deformations. To speed up the computation, a multi-resolution strategy is often used, where the first solution is estimated at a coarse scale (less pixels to go faster). This first estimate is propagated to the higher scale to be refined and so on to the finest scale. This family of methods include in particular the optical flow, which is a very powerful method for estimating the apparent movement and local displacements of pixels, without prior object detection.

Feature-based algorithms proceed in two phases: 1) a detection of elementary visual features (section 10.4.1), such as key points or contours, and 2) a mapping of the features. This is often the approach used for registering unstabilized images or for tracking objects (section 10.4.4).

10.4. Bioimage analysis

In order to guarantee reliable statistics in life sciences, it is essential to acquire a large number of images, which requires automating image analysis. This also makes it possible to ensure unbiased quantification by a human operator. Many books and much software are dedicated to image analysis in computer vision; however, few are yet focusing on the specificities of bioimages (Miura and Sladoje 2020). Below, we present the main tasks of image analysis, visual feature and object detection, image segmentation and object tracking within the context of bioimaging.

10.4.1. *Feature and object detection*

The detection of elementary visual structures such as points of interest, contours or objects of study (nuclei, cells) in microscopy images is an important step in an image analysis system because it is at the basis of numerous methods for object counting, segmentation and tracking.

The methods mainly rely on conventional computer vision techniques, and more specifically the detection of the following structures:

– Edge detection, based on the computation of the gradient. Edges are useful for characterizing the morphology of objects. It can also be applied in 3D to detect membranes (see Figure 10.6A).

– Ridge detection, based on the computation of the eigenvalues of the Hessian matrix (multidimensional partial second derivative) (Frangi et al. 1998). The ridges correspond to the filaments or fibers often present in images in biology (see Figure 10.6B).

– Keypoint detection, based on the evaluation of the gradient at multiple scales, for instance, with the very popular SIFT algorithm (Lowe 1999). Keypoints are local structures, such as filament ends or junctions. Keypoints are very useful for matching images (sections 10.3.4 and 10.4.4) (see Figure 10.6D).

– Directionality detection, based on the analysis of local structures extracted from the gradient structure tensor matrix (Püspöki et al. 2016). This information is very appropriate for evaluating diffusions, for example (see Figure 10.6E).

– Detection of bright spots, a common operation in bioimaging. A very simple approach consists of using bandpass filters (such as the Difference of Gaussian (DoG) and Laplacian of Gaussian (LoG) (Sage et al. 2005) filters), and then finding the local maxima. If there is a model of the shape of the spots, one detection method is to fit the model to the data, which can have the advantage of accurate, sub-pixel localization. The latter approach is the basis for many SMLM algorithms (see Figure 10.6C).

Because they can be easily controlled, model-based methods are preferable for the detection of these elementary local features that have a very explicit model. For other features, without explicit description, data-driven approaches are more adequate; this is in particular the case for texture analysis (Depeursinge et al. 2017) (see Figure 10.6F).

Deep learning methods are well suited for complex object detection tasks. They strive to find a bounding box around each object of interest in an image (see Figure 10.6G). The most well-known methods for object detection in computer vision include R-CNN, which operates in two steps (Girshick et al. 2014; Girshick 2015). The first step extracts region-proposals, and the second classifies them by refining the positions of the bounding boxes. For its part, the YOLO (You Only Look Once) method makes use of a single convolutional lattice to predict both the bounding boxes and the class probabilities for these boxes (Redmon et al. 2016). While these methods are widely used in computer vision, they have only recently been applied to microscopy images, and mainly for the detection of particles and nuclei (Xiao and Yang 2017; Yuan et al. 2019).

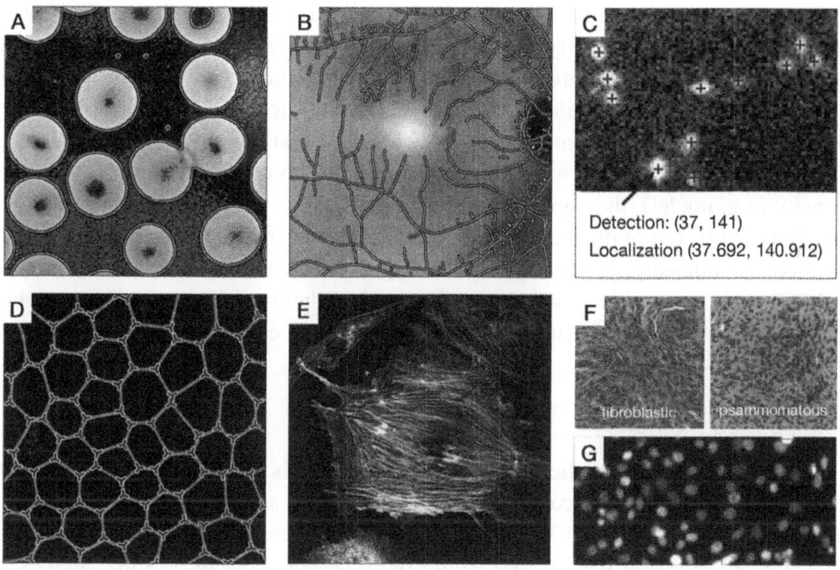

Figure 10.6. *Illustrations of structure and object detection. A. Cell edge detection. B. Ridge detection with the vascular structure of the retina. C. Detection of bright spots (single molecular emitters) using the local maxima of the response of a bandpass filter (DoG). The detected positions are approximated on the pixel grid (in integer value); a localization by fitting a 2D Gaussian function results in a more accurate position (subpixel). D. Three-branch junction detection. E. Detection (color-encoded) of filament directions. F. Examples of two different textures in histopathology images. G. Principle of object detection by bounding box*

10.4.2. *Segmentation*

Segmentation is an often mandatory step in image analysis whose purpose is to partition an image into different regions with similar properties. In other words, segmentation is able to extract objects of interest from the image. Conceptually, its interest lies in obtaining a high-level description of the objects in an image (such as their shape, for example). Segmentation is thus responsible for changing the pixel representation of an image into a representation of its content, such as a series of more relevant objects for the quantification step. This is a challenging task for which a large number of methods have been developed.

Inevitably, no existing segmentation method can be considered generic enough for all applications. They differ according to the imaging modality, the type and shape of the objects of interest, the application and the level of automation.

Pixel-based methods

Basic segmentation methods often include thresholding, which classifies pixels according to their intensity without taking their context into account (see Figure 10.7A). Intensity thresholding is often combined with ad hoc processing solutions to correct unavoidable misclassifications, such as binary morphological operators (erosion, dilation, median) or the watershed that separates objects in contact. While these methods work well for objects that are homogeneous in size and light intensity (e.g. nuclei), they are generally not very robust in bioimaging when a great variability is present in the experimental data. They are indeed prone to experimental variations in intensity. When interaction is possible with the user, semi-automatic methods such as region growing can be used to help with segmentation, which agglomerates pixels of comparable intensity from a source point, called a seed, defined by the user.

Model-based methods

As an alternative to these segmentation methods based on pixel intensity, deformable models, also called *active contours* or *snakes* are widely used in bioimaging (see Figure 10.7B and 10.7C). They were first proposed in 2D by Kass et al. (1988) and generalized to 3D by Terzopoulos et al. (1988). Snakes are semi-automatic methods that offer an excellent trade-off among flexibility, efficiency and interactivity. They enable a manual correction of the segmentation result, which is a significant advantage because automatic segmentation methods rarely achieve an ideal accuracy.

Snakes extract objects of interest in images by deforming their outline towards the edges of the objects of interest by minimizing an energy function whose purpose is to find the optimal configuration of the snake outline. To construct a snake, two elements are needed: 1) a geometric representation of the snake that describes the nature of its outline and determines some geometric properties (e.g. smooth curves, topological properties). There are several families of representations: geodesics, meshes, parametric and subdivisions (Delgado-Gonzalo et al. 2014; Badoual et al. 2016). 2) The appropriate energy function that drives the adjustment of the snake contour to the image data. The choice of this term of energy is crucial because it affects the quality of the result of the segmentation and, as such, is application-dependent. Energy terms can be the attraction to the contour (*edge-based energy*) or the attraction to homogeneous regions (*region-based energy*) (Jacob et al. 2004).

While these methods have many advantages for segmentation in bioimaging, they also present some limitations. First, they require an initial prompt of the snake that can either be done manually or using an auxiliary detection method (see section 10.4.1). Second, the division of objects tends to be excessive when they are very close, which will require a collision detection algorithm. Finally, for images where the contrast between the object of interest and the background is weak, energies that rely simply on intensity information are not very effective. The spatial organization of the pixels (texture) within the object of interest must be incorporated into the energy function (Depeursinge et al. 2017). While some *texture-based energies* have been developed for snakes (Badoual et al. 2019b), data-driven segmentation methods, especially deep learning, excel in this kind of applications when a large amount of example data is available.

Data-driven methods

Data-driven methods usually turn segmentation into a pixel classification problem: each pixel is assigned a label or class, for example, class 0 for the image background, 1 for the nuclei and 2 for the cytoplasm. These supervised machine learning methods are trained from images whose pixels have been previously annotated by field experts.

Among the machine learning methods widely used in bioimaging, we find support vector machines (SVM) (Cortes and Vapnik 1995) or Random Forests (RF) (Ho 1995) particularly used in the Ilastik software (Berg et al. 2019). For these classifiers, the local characteristics of the pixels are calculated from many generic filters such as those presented in section 10.3.1, such as edge detectors, the multiscale Gaussian filter or texture-based filters (Depeursinge et al. 2017).

In deep learning, convolutional neural networks, such as U-Net (Ronneberger et al. 2015; Çiçek et al. 2016) or SegNet (Badrinarayanan et al. 2017), have become very popular because they outperform other segmentation methods of biological images (Ulman et al. 2017). In particular, their ability to automatically determine the right discriminating characteristics of pixels makes them particularly good at classifying pixels. They can be used to segment cells with complex textures, without the need to explicitly define an energy model or function, as shown in Figure 10.7D for the segmentation of HeLa cells. Most of these networks now work with 2D images and also in 3D at the cost of more expensive training. Within this context, these data-driven methods perform semantic segmentation, not instance segmentation. For each pixel, they estimate the probability of belonging to each predicted class (e.g. cell or background). To then achieve a segmentation of instances, the model-based methods mentioned above, such as a combination of thresholding and watershed or active contours, have to be applied to the result.

Instance segmentation has become very popular in computer vision and there are increasingly more pre-trained neural networks for bioimaging applications. This recent trend is evolving fast. At the moment, the best known can be mentioned:

– Cellpose (Schmidt et al. 2018) (see Figure 10.7E and 10.7F) is a model trained to detect cells of different shapes and in different modalities.

– StarDist (Stringer et al. 2021) (see Figure 10.7G and 10.7H) is capable of quickly segmenting thousands of objects with rounded shapes (such as nuclei), even when they can be found in very high density in images.

– YOLO (Redmon et al. 2016) (see Figure 10.7E and 10.7F) is a neural network that places a bounding box around image objects. This network must be re-trained with application-specific data.

– Segment anything for microscopy (Archit et al. 2023) follows the public availability of the segment-anything-model (SAM) which is a "foundation" model trained with 11 million natural (non-microscopy) images annotated with 1 billion masks. Segment anything for microscopy has extended SAM to microscopy data to provide a highly versatile annotation tool.

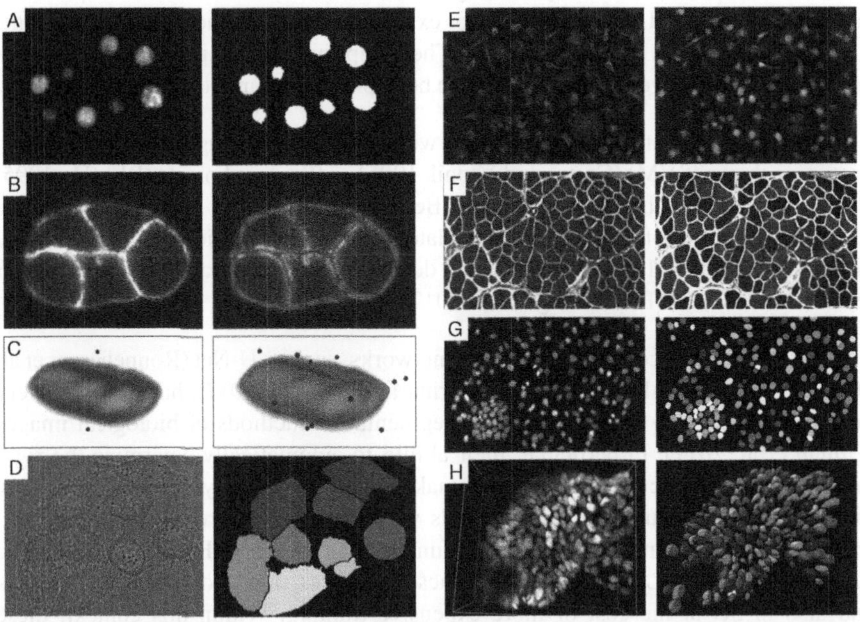

Figure 10.7. *Microscopy image segmentation. A. Nuclei segmentation by thresholding. B. Membrane segmentation of a C elegans embryo with a 2D (ridge-based energy) active contour (Badoual et al. 2019a). Source: R. Jankele and P. Gønczy, EPFL. C. Segmentation of a rat neuron nucleus with a 3D active contour (Badoual et al. 2021). D. HeLa cells acquired in phase contrast segmented with a U-Net-type CNN network. Image of the cell tracking challenge. E. and F. Cell segmentation with Cellpose. Source: http://www.cellpose.org/. G. and H. 2D and 3D dense nuclei segmentation, respectively, with StarDist. Sources: Broad Bioimage Benchmark Collection (Caicedo et al. 2019) and from Schmidt et al. (2018)*

10.4.3. *Image classification*

In bioimaging, the classification of objects of interest is an important step. It can be used for instance to differentiate between living and dead cells, cancer cells or cells during the division cycle, or ribosomes from mitochondria. To this end, there are methods known as image-level classification methods that assign a label to each input image, as opposed to pixel classification methods (seen in section 10.4.2 on segmentation).

It is seldom the case, however, to have to directly classify the entire image, which usually contains a multitude of objects of interest. Instead, sub-images are classified, namely patches of the original image, each comprising a unique object of interest. These patches are obtained by extracting objects in the form of a region of interest (ROI) during the segmentation and/or detection steps introduced earlier.

Most classification methods are based on machine learning. The most conventional approach is the kMeans method, which is an unsupervised method. This is instead referred to as clustering, i.e. the objective consists of separating objects of interest into different classes without attaching a label to them. The kMeans method is applied either directly to a set of images or after using a dimensionality reduction technique (e.g. PCA, T-SNE (Van der Maaten and Hinton 2008)). This method is quick and easy, but highly dependent on initialization and may fail when classes are complex and intertwined. Among the supervised methods widely used in bioimaging, we find the methods mentioned in segmentation for pixel classification, namely SVM and Random Forests, but here they are applied at the image scale and no longer at the pixel level. These classifiers require the extraction of structural features (*image features*) and are trained from labels of classes that must be known. More recently, deep learning methods, especially CNN-based (for instance, VGG16 (Simonyan and Zisserman 2014), ResNet (He et al. 2016)), automatically learn these image features from the data, and have demonstrated their effectiveness for microscopy image classification (Xing et al. 2017; Meijering 2020). The main limitation of these deep learning methods is the need for a large amount of annotated data for training (see section 10.5).

10.4.4. *Tracking of biological objects*

The tracking of cells, nuclei or any other bio-particles is of great importance in bioimaging. The analysis of trajectories in 2D or 3D underlines cell dynamics and interactions between objects of interest.

Tracking is a particularly complex task (Meijering et al. 2006) because there is a great diversity of objects of interest (targets) to be tracked, ranging from the entire animal (e.g. *C elegans*, mice) (Mathis et al. 2018) to the single molecule, including the often very numerous and dense cells in images (see Figure 10.8B). Moreover, their dynamics and cell divisions are all factors that complicate the design of algorithms. In bioimaging, tracking is rarely in real time, unlike computer-based vision applications; it is therefore possible to take the past and the future into account to make algorithms more robust.

The standard approach is to perform a "frame-to-frame" tracking with a spatial detection strategy for objects in each image followed by a time linking of the objects (see Figure 10.8A) (Meijering et al. 2012). The first phase consists of detecting the objects of interest in each of the images with segmentation, model-based or data-driven methods, as previously presented. Each detected object is then described by a vector of characteristics that integrates visual features such as position, size, color or morphology. In a second step, the objects of two consecutive images are linked by similarity and/or proximity by evaluating an energy function. Making the connections in pairs of images only in the sequence has a major disadvantage due to the risk of error propagation; a false detection in one image will have consequences for the rest of the image sequence. In order to make the algorithms more robust, it proves more efficient to account for time persistence, assuming short displacements to integrate past and future information. This is the case of the MHT (Multiple Hypothesis Test (Chenouard et al. 2013) tracking algorithm which is based on probabilistic associations with a few images, or the SpotTracker algorithm which computes the least energy-intensive path of a particle over the entire sequence (see Figure 10.8(D)) (Sage et al. 2005), or still the U-track algorithm which involves a multiple stage linking strategy, by starting by grouping the objects into small segments of trajectories and then combining them step by step (Jaqaman et al. 2008).

Deep learning methods only deal with the spatial segmentation part because the temporal linking step is always more efficient with model-based approaches.

Single molecule trajectography

These tracking methods are also used in the context of single-molecule imaging (SMLM, QDot, sptPALM, uPAINT, etc.) for determining the molecular interactions between the protein of interest and the partner molecules (Sibarita 2014). Using statistical approaches, it is also possible to compute diffusion maps from displacement vectors or from trajectories (Kusumi et al. 2014).

Trajectory analysis

The statistical analysis of trajectories provides relevant information for interpreting molecular interactions, or cell migration speeds. It is possible to calculate a given number of parameters, such as velocities, diffusion and directionality. Mean square displacement (MSD) makes it possible to classify trajectories according to their diffusive behavior, i.e. directional trajectories (superdiffusive), more confined trajectories (subdiffusive) or Brownian motions that are very common in biology, as shown in Figure 10.8C.

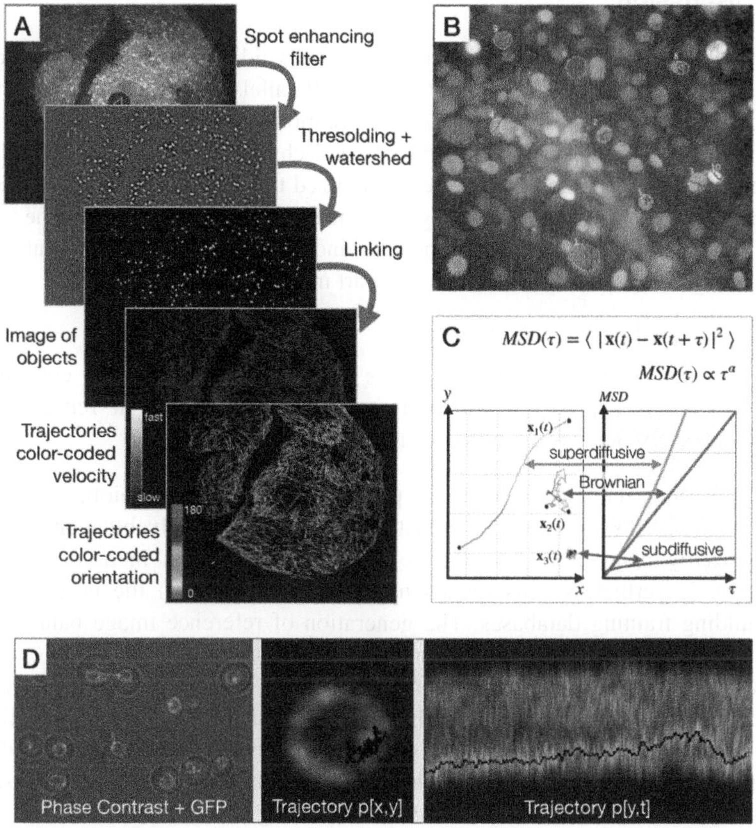

Figure 10.8. *Tracking of biological objects. A. Tracking of microtubule tips in a sequence of 400 images. From top to bottom: example of an image at a given time; image after bandpass filtering that eliminates background and noise; image after thresholding and watershed to isolate objects of interest; temporal linking of all detected objects and view of all trajectories with a color code that indicates the speed; view of all paths with color coding indicating orientation. B. "Image-to-image" segmentation and temporal connection of 10 cells chosen by the user in a very dense medium. Bioluminescence image for studies of the circadian cycle (Sage et al. 2010). C. Three examples of 2D trajectory: $x_1(t)$ with a superdiffusive displacement model, $x_2(t)$ with a diffusive normal displacement model also called the Brownian motion, and $x_3(t)$ with a subdiffusive model, confined to a small space. Left: representation of MSDs corresponding to the three types of trajectories. D. Tracking of a telomere in a cell nucleus. The cell membrane and telomere are labeled in GFP. The trajectory is illustrated by a red line on the planes (X,Y) and (Y,T)*

Trajectory analysis is also used to extract cell lineage including divisions and migrations, for example, for embryogenesis applications.

10.5. Discussion

In this chapter, we have presented the main guidelines for image processing and analysis applied to microscopy images. Parallels are often drawn between two methodological approaches: the conventional model-based approach and the data-driven approach which is based on technologies derived from artificial intelligence. While AI models have revolutionized the world of image analysis, and consequently the world of cellular imaging, fundamental questions are still the topic of debate about the balance between the benefits and risks of AI, the shift to a data-driven paradigm and the expected prospects of "smart microscopy".

Advantages and risks of deep learning

Most image analysis system designers face a dilemma between conventional methods and DL approaches, often using selection criteria that remain largely unmastered or obscure, sometimes driven by biases or current hype.

Deep learning methods have clearly pushed the boundaries of bioimage analysis. They have enabled solving problems that were considered unsolvable with traditional approaches by being able to integrate complex information in a large quantity of images. Nevertheless, this results in significant costs: 1) the cost associated with building training databases. The generation of reference image banks is very labor-intensive in the field of life sciences. Big companies such as Google or Meta have also created large databases of natural images by exploiting the information available on the Internet without much regard for ethical issues and the intellectual property of this information. As mentioned in this chapter, Meta has made available a segment-anything foundation model that can segment microscopy images itself on high-performance machines. 2) The cost of training. Neural networks are indeed becoming larger and larger with millions of parameters to be optimized, which requires significant resources, both computational (e.g. GPU) and human (e.g. design, adjustment of hyper parameters).

DL systems have a "black-box" aspect that scientists hardly appreciate, as they have no proof of convergence and no traceability of errors. On the other hand, model-based systems are well mastered, with results that can be explained relatively to a physical model formulated by engineers; they can be generalized within the framework of established assumptions, and interpreted by humans.

Many applications have combined the two approaches in a hybrid design, taking advantage of both the versatility of DL and the control of physical models. Developers must understand both worlds perfectly to combine them rather than oppose them. They must have sufficient training to be able to distinguish a task that can be performed using a model and thus avoid the risk of using DL when it is not justified.

For a non-expert user, the use of deep learning methods may seem out of reach. For the training phase, not only must large amounts of ground truth data be collected, but a high-performance machine with large memory GPUs is also needed. In addition, it is most of the time necessary to have very good knowledge of programs in Python environments that are known to be difficult for neophytes. Efforts are being made to integrate training programs into more user-friendly environments, for example, with notebooks from ZeroCostDLMic (von Chamier et al. 2021). On the other hand, the use of pre-trained deep learning models is much more accessible, as long as there is a model that is exactly suitable for the intended application. The Bioimage Model Zoo (BMZ) (Ouyang et al. 2022) is a repository of standardized models for bioimaging that attempts to gather many pre-trained models. Deep learning predictions can usually be achieved on standard machines without GPUs. Therefore, ImageJ/Fiji includes a plugin for StarDist, a plugin for Noise2Void and also a deepImageJ plugin that make it possible to execute inferences with all BMZ models (Gómez-de Mariscal et al. 2021).

Shift to data design

DL methods, in supervised mode, require a large number of training images, in particular ground truth images or annotated images. These images must be representative of a given application; in other words, they must cover all the variability of situations. This is particularly problematic in bioimaging, due to the variety of phenotypes, dynamics and interactions of the living environment. In the absence of large reference databases for bioimaging, other strategies must be used:

– large databases of all origins designed for computer vision. The networks pre-trained with these databases can then be fine-tuned from a few application-specific bioimages;

– data annotated manually by experts or, alternatively, by non-experts;

– artificial data generated by a simulator. In fluorescence microscopy, for which image formation models are well mastered, the results are promising;

– microscopy tips for acquiring images in two modes, in high and low qualities (Weigert et al. 2018).

Some initiatives have been taken, for specific applications, to create open image databases (e.g. Kaggle, Grand Challenges), which can be enhanced in a collaborative way.

With the emergence of DL, all the challenges of bioimage analysis have shifted from the design of algorithms based on the development of physical or biological models, to a massive data collection process to build training databases, which represents a major effort by the bioimaging community.

Smart microscopy

A new era has very recently begun with the developments in "smart microscopy" (Carpenter et al. 2023). This approach aims to perform real-time image analysis during the image acquisition process in order to collect only the images necessary for the experiment. This real-time image analysis can be either model-based or data-driven. It therefore regularly provides information for controlling the microscope to adapt the images acquired to the phenotype observed. For example, we can imagine adapting the spatial and temporal resolutions, in real time according to the samples, in order to reduce toxicity by illuminating only what is necessary. Intelligent predictions can also guide acquisition to the right place and at the right time in order to focus only on the event of interest (Mahečić et al. 2022).

Smart microscopy and deep learning combined with model-based image analysis undoubtedly open up new perspectives in bioimaging that could provide intelligent solutions to data flows in an attempt to unravel the secrets of the fabulous cellular machinery.

10.6. References

Anderson, K., Nilsson, T., Fernandez-Rodriguez, J. (2019). Challenges for CLEM from a light microscopy perspective. In *Correlative Imaging: Focusing on the Future*, Verkade, P. and Collinson, L. (eds). Wiley, Hoboken, NJ.

Archit, A., Nair, S., Khalid, N., Hilt, P., Rajashekar, V., Freitag, M., Gupta, S., Dengel, A., Ahmed, S., Pape, C. (2023). Segment anything for microscopy. *bioRxiv*. doi: 10.1101/2023.08.21.554208.

Badoual, A., Schmitter, D., Uhlmann, V., Unser, M. (2016). Multiresolution subdivision snakes. *IEEE Transactions on Image Processing*, 26(3), 1188–1201.

Badoual, A., Galan, A., Sage, D., Unser, M. (2019a). Deforming tessellations for the segmentation of cell aggregates. In *IEEE 16th International Symposium on Biomedical Imaging (ISBI 2019)*, Cambridge, MA.

Badoual, A., Unser, M., Depeursinge, A. (2019b). Texture-driven parametric snakes for semi-automatic image segmentation. *Computer Vision and Image Understanding*, 188, 102793.

Badoual, A., Romani, L., Unser, M. (2021). Active subdivision surfaces for the semiautomatic segmentation of biomedical volumes. *IEEE Transactions on Image Processing*, 30, 5739–5753. doi: 10.1109/TIP.2021.3087947.

Badrinarayanan, V., Kendall, A., Cipolla, R. (2017). Segnet: A deep convolutional encoder-decoder architecture for image segmentation. *IEEE Transactions on Pattern Analysis and Machine Intelligence*, 39(12), 2481–2495.

Batson, J. and Royer, L. (2019). Noise2self: Blind denoising by self-supervision. In *International Conference on Machine Learning*. PMLR, London.

Berg, S., Kutra, D., Kroeger, T., Straehle, C.N., Kausler, B.X., Haubold, C., Schiegg, M., Ales, J., Beier, T., Rudy, M. et al. (2019). Ilastik: Interactive machine learning for (bio) image analysis. *Nature Methods*, 16(12), 1226–1232.

Berlin, S., Carroll, E.C., Newman, Z.L., Okada, H.O., Quinn, C.M., Kallman, B., Rockwell, N.C., Martin, S.S., Lagarias, J.C., Isacoff, E.Y. (2015). Photoactivatable genetically encoded calcium indicators for targeted neuronal imaging. *Nature Methods*, 12(9), 852–858.

Boulanger, J., Kervrann, C., Bouthemy, P., Elbau, P., Sibarita, J.-B., Salamero, J. (2009). Patch-based nonlocal functional for denoising fluorescence microscopy image sequences. *IEEE Transactions on Medical Imaging*, 29(2), 442–454.

Broughton, S.A. and Bryan, K. (2018). *Discrete Fourier Analysis and Wavelets: Applications to Signal and Image Processing*. John Wiley & Sons, New York.

Caicedo, J.C., Goodman, A., Karhohs, K.W., Cimini, B.A., Ackerman, J., Haghighi, M., Heng, C., Becker, T., Doan, M., McQuin, C. et al. (2019). Nucleus segmentation across imaging experiments: The 2018 Data Science Bowl. *Nature Methods*, 16(12), 1247–1253.

Carpenter, A.E., Kamentsky, L., Eliceiri, K.W. (2012). A call for bioimaging software usability. *Nature Methods*, 9(7), 666–670.

Carpenter, A., Cimini, B., Eliceiri, K. (2023). Smart microscopes of the future. *Nature Methods*, 20, 962–964.

von Chamier, L., Laine, R.F., Laine, R.F., Jukkala, J., Spahn, C., Krentzel, D., Krentzel, D., Nehme, E., Lerche, M., Hernández-Pérez, S. et al. (2021). Democratising deep learning for microscopy with ZeroCostDL4Mic. *Nature Communications*, 12(1), 2276–2276.

de Chaumont, F., Dallongeville, S., Chenouard, N., Hervé, N., Pop, S., Provoost, T., Meas-Yedid, V., Pankajakshan, P., Lecomte, T., Le Montagner, Y. et al. (2012). Icy: An open bioimage informatics platform for extended reproducible research. *Nature Methods*, 9(7), 690–696.

Chenouard, N., Bloch, I., Olivo-Marin, J.-C. (2013). Multiple hypothesis tracking for cluttered biological image sequences. *IEEE Transactions on Pattern Analysis and Machine Intelligence*, 35(11), 2736–3750.

Çiçek, Ö., Abdulkadir, A., Lienkamp, S.S., Brox, T., Ronneberger, O. (2016). 3D U-Net: Learning dense volumetric segmentation from sparse annotation. In *International Conference on Medical Image Computing and Computer Assisted Intervention*, Springer, New York.

Cohen, A.R. (2014). Extracting meaning from biological imaging data. *Molecular Biology of the Cell*, 25(22), 3470–3473.

Cortes, C. and Vapnik, V. (1995). Support-vector networks. *Machine Learning*, 20(3), 273–297.

Delgado-Gonzalo, R., Uhlmann, V., Schmitter, D., Unser, M. (2014). Snakes on a plane: A perfect snap for bioimage analysis. *IEEE Signal Processing Magazine*, 32(1), 41–48.

Depeursinge, A., Al-Kadi, O.S., Mitchell, J.R. (2017). *Biomedical Texture Analysis: Fundamentals, Tools and Challenges*. Academic Press, Cambridge, MA.

Dey, N., Blanc-Feraud, L., Zimmer, C., Roux, P., Kam, Z., Olivo-Marin, J.-C., Zerubia, J. (2006). Richardson–Lucy algorithm with total variation regularization for 3D confocal microscope deconvolution. *Microscopy Research and Technique*, 69(4), 260–266.

Ellenberg, J., Swedlow, J.R., Barlow, M., Cook, C.E., Sarkans, U., Patwardhan, A., Brazma, A., Birney, E. (2018). A call for public archives for biological image data. *Nature Methods*, 15(11), 849–854.

Frangi, A.F., Niessen, W.J., Vincken, K.L., Viergever, M.A. (1998). Multiscale vessel enhancement filtering. In *International Conference on Medical Image Computing and Computer-Assisted Intervention*. Springer, New York.

Girshick, R. (2015). Fast R-CNN. In *Proceedings of the IEEE International Conference on Computer Vision*, Cambridge, MA.

Girshick, R., Donahue, J., Darrell, T., Malik, J. (2014). Rich feature hierarchies for accurate object detection and semantic segmentation. In *Proceedings of the IEEE Conference on Computer Vision and Pattern Recognition*, Cambridge, MA.

Gómez-de Mariscal, E., García-López-de Haro, C., Ouyang, W., Donati, L., Lundberg, E., Unser, M., Muñoz-Barrutia, A., Sage, D. (2021). DeepImageJ: A user-friendly environment to run deep learning models in ImageJ. *Nature Methods—Techniques for Life Scientists and Chemists*, 18(10), 1192–1195.

Goodfellow, I., Bengio, Y., Courville, A. (2016). *Deep Learning*. MIT Press, Cambridge, MA.

Gustafsson, M.G. (2000). Surpassing the lateral resolution limit by a factor of two using structured illumination microscopy. *Journal of Microscopy*, 198(2), 82–87.

He, K., Zhang, X., Ren, S., Sun, J. (2016). Deep residual learning for image recognition. In *Proceedings of the IEEE Conference on Computer Vision and Pattern Recognition*, Cambridge, MA.

Ho, T.K. (1995). Random decision forests. In *Proceedings of 3rd International Conference on Document Analysis and Recognition*. IEEE, Cambridge, MA.

Jacob, M., Blu, T., Unser, M. (2004). Efficient energies and algorithms for parametric snakes. *IEEE Transactions on Image Processing*, 13(9), 1231–1244.

Jaqaman, K., Loerke, D., Mettlen, M., Kuwata, H., Grinstein, S., Schmid, S.L., Danuser, G. (2008). Robust single-particle tracking in live-cell time-lapse sequences. *Nature Methods*, 5(8), 695–702.

Kass, M., Witkin, A., Terzopoulos, D. (1988). Snakes: Active contour models. *International Journal of Computer Vision*, 1(4), 321–331.

Kefer, P., Iqbal, F., Locatelli, M., Lawrimore, J., Zhang, M., Bloom, K., Bonin, K., Vidi, P.-A., Liu, J. (2021). Performance of deep learning restoration methods for the extraction of particle dynamics in noisy microscopy image sequences. *Molecular Biology of the Cell*, 32(9), 903–914.

Kirshner, H., Aguet, F., Sage, D., Unser, M. (2013). 3-D PSF fitting for fluorescence microscopy: Implementation and localization application. *Journal of Microscopy*, 249(1), 13–25.

Krull, A., Buchholz, T.-O., Jug, F. (2019). Noise2void-learning denoising from single noisy images. In *Proceedings of the IEEE/CVF Conference on Computer Vision and Pattern Recognition*. IEEE, Cambridge, MA.

Kusumi, A., Tsunoyama, T.-A., Hirosawa, K., Kasai, R., Fujiwara, T. (2014). Tracking single molecules at work in living cells. *Nature Chemical Biology*, 10, 524–532.

Laine, R.F., Jacquemet, G., Krull, A. (2021). Imaging in focus: An introduction to denoising bioimages in the era of deep learning. *The International Journal of Biochemistry & Cell Biology*, 140, 106077.

Lehtinen, J., Munkberg, J., Hasselgren, J., Laine, S., Karras, T., Aittala, M., Aila, T. (2018). Noise2noise: Learning image restoration without clean data. *arXiv*, 1803.04189.

Lowe, D.G. (1999). Object recognition from local scale-invariant features. In *Proceedings of the Seventh IEEE International Conference on Computer Vision*, Cambridge, MA.

Lucas, A.M., Ryder, P.V., Li, B., Cimini, B.A., Eliceiri, K.W., Carpenter, A.E. (2021). Open-source deep-learning software for bioimage segmentation. *Molecular Biology of the Cell*, 32(9), 823–829.

Luisier, F., Vonesch, C., Blu, T., Unser, M. (2010). Fast interscale wavelet denoising of Poisson-corrupted images. *Signal Processing*, 90(2), 415–427.

Mahečić, D., Stepp, W., Zhang, C., Griffie, J., Weigert, M., Manley, S. (2022). Event-driven acquisition for content-enriched microscopy. *Nature Methods*, 19(10), 1262–1267.

Mathis, A., Mamidanna, P., Cury, K.M., Abe, T., Murthy, V.N., Mathis, M.W., Bethge, M. (2018). DeepLabCut: Markerless pose estimation of user-defined body parts with deep learning. *Nature Neuroscience*, 21(9), 1281–1289.

Meijering, E. (2020). A bird's-eye view of deep learning in bioimage analysis. *Computational and Structural Biotechnology Journal*, 18, 2312.

Meijering, E., Smal, I., Danuser, G. (2006). Tracking in molecular bioimaging. *IEEE Signal Processing Magazine*, 23(3), 46–53.

Meijering, E., Dzyubachyk, O., Smal, I. (2012). Methods for cell and particle tracking. *Methods in Enzymology*, 504, 183–200.

Meijering, E., Carpenter, A.E., Peng, H., Hamprecht, F.A., Olivo-Marin, J.-C. (2016). Imagining the future of bioimage analysis. *Nature Biotechnology*, 34(12), 1250–1255.

Meiniel, W., Olivo-Marin, J.-C., Angelini, E.D. (2018). Denoising of microscopy images: A review of the state-of-the-art, and a new sparsity-based method. *IEEE Transactions on Image Processing*, 27(8), 3842–3856.

Miura, K. and Sladoje, N. (2020). *Bioimage Data Analysis Workflows*. Springer, Berlin.

Myers, G. (2012). Why bioimage informatics matters. *Nature Methods*, 9(7), 659–660.

Ouyang, W. and Zimmer, C. (2017). The imaging tsunami: Computational opportunities and challenges. *Current Opinion in Systems Biology*, 4(9), 105–113.

Ouyang, W., Beuttenmueller, F., de Mariscal, E.G., Pape, C., Burke, T., de Haro, C.G.-L., Russell, C., Moya-Sans, L., de-la Torre-Gutiérrez, C., Schmidt, D. et al. (2022). Bioimage model zoo: A community-driven resource for accessible deep learning in bioimage analysis. *bioRxiv*. doi: 10.1101/2022.06.07.495102.

Püspöki, Z., Storath, M., Sage, D., Unser, M. (2016). Transforms and operators for directional bioimage analysis: A survey. In *Focus on Bio-Image Informatics*, De Vos, W.H., Munck, S., Timmermans, J.-P. (eds). Springer, Cham, 69–93.

Redmon, J., Divvala, S., Girshick, R., Farhadi, A. (2016). You only look once: Unified, real-time object detection. In *Proceedings of the IEEE Conference on Computer Vision and Pattern Recognition*, Cambridge, MA.

Richardson, W.H. (1972). Bayesian-based iterative method of image restoration. *JoSA*, 62(1), 55–59.

Ronneberger, O., Fischer, P., Brox, T. (2015). U-Net: Convolutional networks for biomedical image segmentation. In *International Conference on Medical Image Computing and Computer Assisted Intervention*. Springer, New York.

Russ, J.C. (2002). *The Image Processing Handbook*. CRC Press, Boca Raton, FL.

Sage, D., Neumann, F.R., Hediger, F., Gasser, S.M., Unser, M. (2005). Automatic tracking of individual fluorescence particles: Application to the study of chromosome dynamics. *IEEE Transactions on Image Processing*, 14(9), 1372–1383.

Sage, D., Unser, M., Salmon, P., Dibner, C. (2010). A software solution for recording circadian oscillator features in time-lapse live cell microscopy. *Cell division*, 5(1), 1–9.

Sage, D., Donati, L., Soulez, F., Fortun, D., Schmit, G., Seitz, A., Guiet, R., Vonesch, C., Unser, M. (2017). DeconvolutionLab2: An open-source software for deconvolution microscopy. *Methods*, 115, 28–41.

Sage, D., Pham, T.-A., Babcock, H., Lukes, T., Pengo, T., Chao, J., Velmurugan, R., Herbert, A., Agrawal, A., Colabrese, S. et al. (2019). Super-resolution fight club: Assessment of 2D and 3D single-molecule localization microscopy software. *Nature Methods*, 16(5), 387–395.

Sarder, P. and Nehorai, A. (2006). Deconvolution methods for 3-D fluorescence microscopy images. *IEEE Signal Processing Magazine*, 23(3), 32–45.

Schindelin, J., Rueden, C.T., Hiner, M.C., Eliceiri, K.W. (2015). The ImageJ ecosystem: An open platform for biomedical image analysis. *Molecular Reproduction and Development*, 82(7–8), 518–529.

Schmidt, U., Weigert, M., Broaddus, C., Myers, G. (2018). Cell detection with star-convex polygons. In *International Conference on Medical Image Computing and Computer Assisted Intervention*. Springer, New York.

Sibarita, J.-B. (2014). High-density single-particle tracking: Quantifying molecule organization and dynamics at the nanoscale. *Histochemistry and Cell Biology*, 141.

Simonyan, K. and Zisserman, A. (2014). Very deep convolutional networks for large-scale image recognition. *arXiv*, 1409.1556.

Soille, P. (2013). *Morphological Image Analysis: Principles and Applications*. Springer Science & Business Media, Berlin.

Stringer, C., Wang, T., Michaelos, M., Pachitariu, M. (2021). Cellpose: A generalist algorithm for cellular segmentation. *Nature Methods*, 18(1), 100–106.

Terzopoulos, D., Witkin, A., Kass, M. (1988). Constraints on deformable models: Recovering 3D shape and nonrigid motion. *Artificial Intelligence*, 36(1), 91–123.

Tikhonov, A.N. (1963). Solution of incorrectly formulated problems and the regularization method. *Soviet Math Dokl*, 4, 1035–1038.

Ulman, V., Maška, M., Magnusson, K.E., Ronneberger, O., Haubold, C., Harder, N., Matula, P., Matula, P., Svoboda, D., Radojevic, M. et al. (2017). An objective comparison of cell-tracking algorithms. *Nature Methods*, 14(12), 1141–1152.

Valli, J., Garcia-Burgos, A., Rooney, L.M., de Melo E Oliveira, B.V., Duncan, R.R., Rickman, C. (2021). Seeing beyond the limit: A guide to choosing the right super resolution microscopy technique. *Journal of Biological Chemistry*, 297(1), 100791.

Van der Maaten, L. and Hinton, G. (2008). Visualizing data using t-SNE. *Journal of Machine Learning Research*, 9(11), 2579–2605.

Weigert, M., Schmidt, U., Boothe, T., Müller, A., Dibrov, A., Jain, A., Wilhelm, B., Schmidt, D., Broaddus, C., Culley, S. et al. (2018). Content-aware image restoration: Pushing the limits of fluorescence microscopy. *Nature Methods*, 15(12), 1090–1097.

Xiao, Y. and Yang, G. (2017). A fast method for particle picking in cryo-electron micrographs based on fast R-CNN. *AIP Conference Proceedings*, 1836, 020080.

Xing, F., Xie, Y., Su, H., Liu, F., Yang, L. (2017). Deep learning in microscopy image analysis: A survey. *IEEE Transactions on Neural Networks and Learning Systems*, 29(10), 4550–4568.

Yuan, P., Rezvan, A., Li, X., Varadarajan, N., Van Nguyen, H. (2019). Phasetime: Deep learning approach to detect nuclei in time lapse phase images. *Journal of Clinical Medicine*, 8(8), 1159.

Zimmermann, T., Rietdorf, J., Pepperkok, R. (2003). Spectral imaging and its applications in live cell microscopy. *FEBS Letters*, 546(1), 87–92.

List of Authors

Anaïs BADOUAL
SAIRPICO
Centre INRIA de l'Université de Rennes
France

Pierre BON
XLIM
UMR 7252, CNRS
University of Limoges
France

Dominique BOURGEOIS
Institut de Biologie Structurale
CEA, CNRS
Université Grenoble Alpes
France

Laurent COGNET
Laboratoire Photonique, Numérique et Nanosciences
Institut d'Optique Graduate School
UMR 5298, CNRS
University of Bordeaux
France

Tom DELAIRE
Interdisciplinary Institute for Neuroscience
UMR 5297, CNRS
University of Bordeaux
France
and
Department of Experimental Medical Science
and
Lund Bioimaging Center
Lund University
Sweden

Marina S. DIETZ
Single-Molecule Biophysics
Institute of Physical and Theoretical Chemistry
Goethe University Frankfurt
Germany

Guillaume DUPUIS
Institut des Sciences Moléculaires d'Orsay
Université Paris-Saclay
France

Cyril FAVARD
Infectious Disease Research Institute of Montpellier
IRIM, CNRS
France

Alexandra FRAGOLA
Institut des Sciences Moléculaires d'Orsay
Département de Physique
UMR 8214, CNRS
Université Paris Saclay
France

Rémi GALLAND
Interdisciplinary Institute for Neuroscience
UMR 5297, CNRS
University of Bordeaux
France

Mike HEILEMANN
Single-Molecule Biophysics
Institute of Physical and Theoretical Chemistry
Goethe University Frankfurt
Germany

Veera Venkata Gopala Krishna INAVALLI
Center for Cancer Immunology
University of Southampton
United Kingdom

Antony LEE
Laboratoire Photonique, Numérique et Nanosciences
Institut d'Optique Graduate School
UMR 5298, CNRS
University of Bordeaux
and
Physique des Cellules et Cancer
Institut Curie
UMR 168, CNRS
Université PSL
Sorbonne Université
France

Urs Valentin NÄGERL
Department of Anatomy and Cell Biology
Universitätsmedizin Göttingen
Georg-August-Universität
Germany

Daniel SAGE
Biomedical Imaging Group and Center for Imaging
École Polytechnique Fédérale de Lausanne (EPFL)
Switzerland

Jean-Baptiste SIBARITA
Interdisciplinary Institute for Neuroscience
UMR 5297, CNRS
University of Bordeaux
France

Jip WULFFELÉ
Institut de Biologie Structurale
CEA, CNRS
Université Grenoble Alpes
France

Index

Printed and bound by CPI Group (UK) Ltd, Croydon, CR0 4YY

24/08/2026

14947833-0001